Dialogues and Games of Logic

Volume 6

Meaning and Intentionality

A Dialogical Approach

Volume 1
How to Play Dialogues: An Introduction to Dialogical Logic
Juan Redmond and Matthieu Fontaine

Volume 2
Dialogues as a Dynamic Framework for Logic
Helge Rückert

Volume 3
Logic of Knowledge. Theory and Applictions
Cristina Barés Gómez, Sébastien Magnier, and Francisco J. Salguero, eds.

Volume 4
Hintikka's Take on Realism and the Constructivist Challenge
Radmila Jovanović

Volume 5
The Dialogue between Sciences, Philosophy and Engineering. New Historical and Epistemological Insights. Homage to Gottfried W. Leibniz 1646-1716
Raffaele Pisano, Michel Fichant, Paolo Bussotti and Agamenon R. E. Oliveira, eds. With a foreword by Eberhard Knobloch

Volume 6
Meaning and Intentionality. A Dialogical Approach
Mohammad Shafiei

Dialogues and Games of Logic Series Editors
Shahid Rahman
Nicolas Clerbout
Matthieu Fontaine

shahid.rahman@univ-lille3.fr

Meaning and Intentionality
A Dialogical Approach

Mohammad Shafiei

© Individual author and College Publications 2018.
All rights reserved.

ISBN 978-1-84890-259-6

College Publications
Scientific Director: Dov Gabbay
Managing Director: Jane Spurr
Department of Informatics
King's College London, Strand, London WC2R 2LS, UK

www.collegepublications.co.uk

Original cover design by Laraine Welch
Printed by Lightning Source, Milton Keynes, UK

All rights reserved. No part of this publication may be reproduced, stored in a retrieval system or transmitted in any form, or by any means, electronic, mechanical, photocopying, recording or otherwise without prior permission, in writing, from the publisher.

To Arezoo

Acknowledgments

I wish to thank Dr. Mark van Atten, to whom I owe much in carrying out this work, for his sympathetic, practical advice and support as well as for his precious and precise remarks, both in regard to the content and the form, and the fruitful and clarifying conversations concerning the topics discussed here. I express my gratitude to him.

Also, I thank Prof. Shahid Rahman, for his constructive suggestions about the content of the thesis—specially concerning the last two chapters—and also for all his kind supports during the work on this project. I am grateful to him for his kindness and trust.

I am really grateful to the jury members of my thesis, professor Gerhard Heinzmann, professor Mathieu Marion and Dr. Carlos Lobo. This book in its final version has profited from the valuable remarks offered by professor Marion and Dr. Lobo. I really appreciate it.

During the work on this thesis I benefited from some helpful and inspiring conversations and correspondence with philosophers Philipp Berghofer, Christian Beyer, Nicolas Clerbout. Mirja Hartimo, Manuel Gustavo Issac, Hana Karpenko, Marco Panza and Göran Sundholm. I sincerely thank them. I must also thank all the members of *L'École doctorale de philosophie* of the university of Paris 1, Panthéon-Sorbonne, and the Institute for the History and Philosophy of Science and Technology (IHPST) for the practical supports as well as for providing a vivid and dynamic environment.

Lastly, I thank my parents and parents-in-law for their support and encouragement. I am indebted to them. My special thanks are due to my wife, Arezoo Ashjaei Khameneh, to whom I dedicate this work.

Contents

Acknowledgments — vii

Introduction — 1

1 **The Possibility of Inner Dialogue and its Primordiality** — 7
  1.1 Introduction — 7
  1.2 Wittgenstein's Argument Against the Possibility of Private Language — 10
  1.3 Husserl and Inner Language — 27
    1.3.1 Immateriality of language — 30
    1.3.2 Uncommunicative expressions — 32
    1.3.3 Prayer, inwardness and self-directing — 33
  1.4 Descriptive Phenomenology of Expression — 35
    1.4.1 Language and the distinction between meaning and sense — 35
    1.4.2 Expression and indication — 50
    1.4.3 Three accounts of language — 57
  1.5 Constitutive Phenomenology of Expression — 62
  1.6 Inner Alterity of the Ego — 70
  1.7 Reply to Wittgenstei — 75
    1.7.1 Rule-Following argument and the place of the Meaning — 77

		1.7.2	Substitution argument and the function of language	81
		1.7.3	What is the private?	83
		1.7.4	Stage-setting argument and the distinction between expression and indication	85
	1.8	Conclusion		87

2 Meaning and the Unintuitive   89

     2.1   Introduction, Statement of the Problem 89
     2.2   A Place for the Unintuitive 92
         2.2.1   The difference between the primordial and the non-primordial 92
         2.2.2   The theory of signitive intention 96
         2.2.3   Signitive intention and meaning 102
     2.3   Categorial Synthesis 106
         2.3.1   Signifying feature of meaning 106
         2.3.2   Differences of the viewpoint defended here from any kind of formalism 109
         2.3.3   The interplay between construction and intuition 111
     2.4   Categorial Intuition 118
         2.4.1   Intuition of the categorial form 121
         2.4.2   Ambiguity of the expression "object of intention" 124
         2.4.3   The realm of logos and the act of categorial synthesis 126
         2.4.4   Categorial synthesis versus aesthetic synthesis   128
         2.4.5   Three possible meanings of "categorial intention" 133
         2.4.6   The fulfillment of the intention toward the categorial form 135
     2.5   Complex Meanings 139
         2.5.1   The proposition and the state of affairs . . . 139

CONTENTS xi

		2.5.2	Propositional truth and falsity	143
	2.6	Proposition, Judgment and Belief		145
		2.6.1	Knowledge and Judgment	147
		2.6.2	Judgment and belief	150
		2.6.3	Encounter with some well-known theses	153
		2.6.4	Does belief need evidence?	157
		2.6.5	Judgment without evidence	159
		2.6.6	Summary of the section	161
	2.7	Husserl's Idea of Pure Logic		162
		2.7.1	Semiotic theory of unintuitive thought	162
		2.7.2	Intentional theory of the unintuitive thought	166
		2.7.3	Three levels of formal logic	172
		2.7.4	Normative standpoint toward logic	174
3	**Phenomenology and Dialogical Semantics**			**179**
	3.1	Links between the Phenomenology and Dialogical Approach		179
		3.1.1	Stephen Strasser's book: *The Idea of Dialogal Phenomenology*	180
	3.2	A Short Introduction		197
	3.3	Phenomenology of Dialogical Logic		201
		3.3.1	The transcendental status of dialogue: the idea of modification	201
		3.3.2	Play level vs strategy level as related to consequence-logic vs truth-logic	202
		3.3.3	Dialogical framework on the meaning of logical connectives	207
		3.3.4	Proof or evidence?	209
	3.4	Dummett and Verficationism		215
		3.4.1	Dummett on the philosophical importance of intuitionism	215
		3.4.2	From the verificationist theory of truth to the verificationist theory of meaning	216

		3.4.3	Dummett's argument as it relies on the notion of learning	223
		3.4.4	Transcendental phenomenology on realism vs anti-realism	227
		3.4.5	Against linguisticism	230
		3.4.6	Summary of the section	231
	3.5	Logical Monism		235
4	**Dialogical Apophantics: Formal Analyses**			**237**
	4.1	Ideas toward a Unified Logic		237
		4.1.1	Some significant insights from different logics	239
	4.2	Negation		243
		4.2.1	A review on the role of negative judgments .	244
		4.2.2	Kinds of propositional negations	257
		4.2.3	To examine some suggestions for logics with two negations	263
		4.2.4	Strong negation; and the case of *ex falso* . .	273
		4.2.5	Weak negation; to introduce a new rule . . .	284
		4.2.6	A logic with two negations	287
	4.3	Necessity and Strict Implication		296
		4.3.1	Thesis: the Leibnizian account of necessity .	298
		4.3.2	Strict implication	303
		4.3.3	A Rule for the canonical truths and a logical definition for necessity	314
		4.3.4	Analysis: modality and dialogue	323
	4.4	Conclusion		325

Bibliography   **327**

Index of citations   **340**

Introduction

The objective of the present work is to develop a theory of meaning based on the method of transcendental phenomenology. Our point of departure will be the idea of intentionality, and we will study the constitution of the meaning and will analyze the intentional acts functioning in such a constitution. There are some well known, rival theories of meaning in the literature. Most notably, we can mention the referential theory, and in relation to it the truth-conditional and the model-theoretic approaches, and the use theory of meaning, and in relation to this latter inferentialism as well as verificationstic and proof-theoretic accounts. Of course some of these positions have some concurrences. The phenomenological theory of meaning, as it will be investigated, disagrees with each of the mentioned positions in the basic points of the subject. Thus, some parts of the present project are dedicated to the critical discussions. However, the main aim of such critical discussions would be to help better to clarify the peculiarities of the phenomenological theory at work.

One important issue for any theory of meaning is the explanation of the meaning of logical—and also mathematical—constants. In accordance with the intentional theory, our thesis is that to use the idea of dialogical interaction, as a basic theme in the act of logical reasoning, will be helpful to explicate the meaning of logical constants. In this way, it will be shown that the dialogical logic—introduced by Paul Lorenzen and Kuno Lorenz and then developed by Shahid Rahman and others—is an appropriate framework to carry out the phenomenological investigations in the field of logic, in both meaning clarification and rigorous formalization. The idea is to a appeal to the dialogical semantics in order to develop a comprehensive, phenomenological theory of meaning as well as to advance in realization of the phenomenological idea of pure logic and formal apophantics.

The principal method of the present work is the transcendental phenomenology, as introduced by Edmund Husserl. As far as concerned logical studies we will also use the dialogical semantics. Thus, in a part of the projects certain investigations will be carried out to demonstrate that the phenomenology would grant the foundations of the dialogical method; and this will offer a somehow novel interpretation of the dialogical logic. In regard to logic itself, we will prefer the intutionistic approach, developed by Brouwer and Heyting, whose affinity to the phenomenology is already explored, for instance by the works of van Atten, and we take it for granted. Nevertheless, we will also deal with some aspects of the classical logic within the relevant discussions. However, the main concern of this writing is the meaning and the themes related to it, and we will engage in investigations pertained to logic in so far as they concern the philosophical considerations about the meaning of the logical issues.

Here I just mention the main theses, without further argument, that will be argued for and elaborated in the body of this work according to the order that is explained in the synopsis at the end of this introduction.

Meaning is constituted in the act of expression. Expression and the expressional experience are among peculiar capabilities of the transcendental ego. Not every experience is expressional though an expression may afterwards supervene to it. However, the expressional moment is an essential feature of most of experiences so that in such experiences a meaning-function is intrinsic to the intentionality at work. There are various kinds of expressionoal experiences and hence various categories of expression. Most importantly we have, pre-predicative experiences, thus pre-predicative expressions and meanings, predicative experiences, thus propositional expressions, and post-predicative experiences. Meaning, first of all, is brought about by means of the primordial expression, that is through the expression an ideal meaning supervenes to the content of intuition (of a positional experience). Further meanings can be constructed by means of the categorial synthesis over the already accessible meanings. The constructed meanings, themselves as ideal moments, may serve in constituting further intentions—such intentions are among signitive intentions since they are at work in the absence of the intuitive content to be corresponding to them. However, such intentions may make possible some further experiences and themselves may be fulfilled or not in the course of subsequence activities and encounters with the objectivities.

Meaning is a primitive notion and it can not be reduced to some other notions. The proper aim of a theory of meaning should be to explain its constitution, the ideal relation among the meanings, and the relation of the acts concerning the meaning with other related activities of the transcendental subjectivity, and inter-subjectivity as well. Meaning is an ideal entity which by itself is prior to the sign, verbal or written, assigned to it. Therefore, the rules governing the relation among signs should not be confused with the eidetic rules governing the realm of the meaning, or the realm of *logos* as Husserl calls it. Of course, in some cases the former may reflect the latter and in this respect the studies belonging to the former may be helpful to an investigation concerning the theory of meaning.

The view that the ultimate unit of meaningfulness is proposition should be set aside. Meaning is of various categories one of which is proposition. As we have pre-predicative experiences, we have sub-propositional meanings which are in some cases are prior to the proposition and themselves may be employed to compose a proposition via categorial synthesis. On the other hand, the idea that the beginning of the meaning constitution should be sought for in the interjection is refuted as well by the phenomenology. Meaning of different categories may be constituted by one of these ways: either by means of a primordial expression when the ego possesses an intuition, which is of a categorial articulation, or by means of the categorial synthesis (or analysis) over meanings. Notice that in the former case we implied that the categorial moment may be found in the intuition itself, that is to say that the categorial intuition should be possible. This is indeed true according to the phenomenology: not every intuition is sensuous but there are categorial intuitions. This latter is tightly connected to the idea of *essence seeing* or *ideation*.

In logic we primarily deal with the judgment whose ideal form is proposition. The judgment is brought about when a propositional intention is fulfilled. What fulfills a judgment is called *evidence*. Evidence may be direct, i.e. intuitive, or indirect, i.e. signitive. Any kind of reasoning is to search for an indirect fulfillment for a proposition (regardless of whether or not there is a direct piece of evidence for it), so that a judgment would be made as the consequence. Therefore, in logic we deal with the modifications of the judgment and correspondingly with the propositional constants. The meanings of such constants should be explained in terms of the *intentions* at work, independently of the moments of fulfillment— while the issues about truth, validity and thus *proof* stand in the side of fulfillment. A proper meaning-explanation of the logical constants should deal with the intentionalities and the interactions which take place in the argumentation; and the dialogical framework is quite appropriate in this regard.

The present work consists of four chapters . The first chapter contains some studies about the phenomenology of the language and the possibility of the inner dialogue. The inner dialogue is an important theme for our project, since we wish to use dialogical method alongside with the transcendental phenomenology so that we have to show that the dialogue as such does not depend on the social issues and will not be suspended by means of the transcendental *epoche*. That is to say, the notion of dialogue recalled with the dialogical logic, we are to argue, can be considered as an inner dialogue between the ego and its modifications, and will not render our account of logic to a kind of socialogism. Since there is a famous argument, stated by Wittgenstein, against the possibility of solitary use of language, we examine this argument and discuss a phenomenological response to it, in sections 2 and 7. In section 4, we will try to clarify the notion of "meaning" and resolve some ambiguities around it. Section 5 is dedication to a analysis of the experience involving the primordial expression. Section 6 contains a phenomenological study of the inner alterity as it is at work in the inner dialogue.

The core discussions of the presented theory of meaning are placed in chapter 2. This chapter includes the arguments about the non-primordial expression, or expression in the absence of intuition, thus the case of empty-modification and also that of categorial synthesis (Sections 2 and 3). Some important problems around the idea of categorial intuition will be discussed in section 4. Since, proposition is of a special interest in the philosophical studies about logic, we will discuss this case in more details, and the section 5 and 6 are dedicated to the studies about proposition, judgment and evidence. At the end of this chapter we will summarize the phenomenological account of meaning and accordingly the idea of pure logic as an *a priori* science. As I said, chapter 2 contains the basic elements of the introduced theory; however, the preliminarily considerations stated in chapter 1, specially in sec. 1.4.1, are of high relevance.

In chapter 3, we will discuss the possible connections between the phenomenology and the dialogical method, and dialogical semantics in particular. This chapter also contains some critical discussions against use-theoretic, and specially Dummett's verificationist, account of meaning. These critical treatments are supposed to clarify the phenomenological attitude toward the dialogical method and saves it from some presuppositions, specially those connecting dialogical method to the linguistic pragmatism.

In chapter 4, we will use the conceptual tools developed in the previous chapters in order to explain the meaning of some logical constants, as examples, to show the feasibility of developing a dialogical apophantics as connected to the phenomenological idea of pure logic. This chapter contains some novel ideas to distinguish two kinds of negation, and introduce semantics rules for each of them, and also to offer an interpretation of necessity which is apart from both model-theoretic, possible worlds semantics and proof-theoretics definition. The aim of this chapter is only to carry out some meaning explanations and show their effects to the formalization of logic without going through the formal issues.[1]

---

[1] The axiomatization of the logic, so introduced, and to show completeness results for it are beyond the scope of the present book.

Chapter 1

The Possibility of Inner Dialogue and its Primordiality

1.1 Introduction

Is it possible to speak to oneself? On one hand, there is obviously the prevalent phenomenon of inner speech. On the other hand, there are some doubts about the genuineness of this phenomenon, doubts which assert that the inner dialogue is either completely senseless or merely a reflexion or specter of the communicative speech in one's mind. Such doubts seem plausible because language is commonly taken (or tacitly defined) to be a tool for communication, then since there is no interpersonal communication in solitary life it seems that there is no original room for inner speech. Of course there are some more profound and complicated arguments on this issue in philosophical literature. We will study, , as the starting point of our investigation, one of the most influential arguments which refute the originality of inner dialogue, namely Wittgenstein's private language argument.

There is no doubt that we speak to ourselves in solitude. The question is that whether this phenomenon has a philosophical importance or whether there is something intellectually special in inner dialogue which is not in outer one. If we can find a genuine foundation for inner use of language and overcome the doubts about it, then it will be possible to speak of, and investigate, the peculiarities of inner dialogue and its probably special contents. Furthermore, we would be able to investigate the elements of the act of expressing which can thus be grasped more purely than while considered in the communicative context where there exist some other, themselves blurred, elements.

Our method in this investigation is phenomenology and we will use its basic ideas to clarify the various aspects of the alleged inner language; and we will search for the probable foundations of the possibility of inner dialogue and will analyze its constituents by the method of transcendental phenomenology. Moreover, we are to show how Husserl's phenomenology would deal with Wittgenstein's argument against private language.

Derrida (Derrida, 1973, p.43) claims that the possibility of the *epoché* is equivalent to the possibility of monologue. If Derrida was justified in his claim then if Wittgenstein's argument about the impossibility of original inner speech is valid, this entails the impossibility of the transcendental phenomenology. On the other hand if there can be found strong bases for inner dialogue then the whole project of linguistic turn will turn out to be inappropriate (or pointless); because if we can demonstrate that inner dialogue is primordial in a way that it can be accomplished without any prior dependence on outer dialogue it means that the outer, concrete language, i.e. the ordinary language, is not a necessary condition for the possessing concepts and performing intellectual activity. Accordingly, language can not be considered as the original home of the problems that pertain to the theory knowledge, i.e. problems of meaning, truth, certainty, etc.

Therefore, one will not be so unjustified if asserts that the inner

1.1. INTRODUCTION

dialogue is the point in which many debates and controversies in philosophy meet each other. Since we aim in general to approach to phenomenology of dialogue, to deal with phenomenology of inner dialogue at the first step will be quite helpful; by dealing with it, we can also declare our position in respect with some controversial issues from the very beginning.

In the following, I will first study Wittgenstein's argument and his reasons to refute private language. Then, since our framework is phenomenology, it will be good to have a brief review of what Husserl has said about inner dialogue, in order to see the initial disagreement and the motivation to carry out a critical discussion on the topic. After that, in the main part of this writing we apply phenomenological method to clarify the status of inner dialogue and that whether or not it is transcendentally possible[1]. Here I will study the roots of possibility of inner dialogue and the question that whether these roots are so fundamental that allow our investigation to be considered as an investigation about the origin of language. Before final remarks I will comeback to Wittgenstein's argument and examine it according to the consequences of our investigation. At the end, I will review the conclusion and restate the theses which in this writing will have been defended.

In what follows I will frequently use the terms "inner language", "inner dialogue" and "private language". I shall determine their differences as following: By "inner language" we mean a language which can be originated in solitude, i.e. by a person considered in isolation, thus this language is "inner" because it is not originally created for external uses, namely uses in community. The difference

---

[1] I used here the important notion of "transcendental possibility" which is introduced by van Atten (van Atten, 2001):

> An object is transcendentally possible exactly if it is conceptually possible and moreover can (ideally) be constituted with full evidence.

of the term inner language with the term inner dialogue is that in the latter we consider the two-sided-ness of language using, instead of mere (alleged) classifying role of inner language; it means we need inner alterity beside inner language in order to have inner dialogue. "Private language" in a sense means a language which is created and used only by one person and no one else could be able to use or understand it. Wittgenstein's argument, as we will see, is not restricted to refute only this notion, but his argument is against the possibility of originating a language in solitude, so it is against inner language. That is, in the context of Wittgenstein's discussion "private" is not only used for the language that is supposed to be only privately used but also for the language that is supposed to privately created whether or not it is used publicly thereafter. In short, in Wittgenstein's argument "private language" and "inner language" is used interchangeably.

All of these are different from "mental speech". The latter is not controversial. One who rejects private language and thus rejects the genuineness of inner dialogue can acknowledge mental speech, saying that it is not a genuine dialogue but a mere imagination of dialogue, or that it is an undue and futile employing of language. We should emphasize that our theme is not mental speech which is an undoubted phenomenon. Our theme is inner dialogue as considered away from mere being an effect of outer language and as (if possible) having its own peculiarities.

1.2 Wittgenstein's Argument Against the Possibility of Private Language

In some passages of his *Philosophical Investigation*, Wittgenstein discusses to refute the possibility of private language. According to Wittgenstein, it is meaningless to speak about language unless in the context of the communication in a community. Now let us see

1.2. PRIVATE LANGUAGE ARGUMENT

that how his argument is to work. There is no consensus on reading of Wittgenstein's argument. However, his argument about private language consists of certain elements. In the following I first present a thematic report of how Wittgenstein's discussion goes. Then, I will extract the consisting elements and reconstruct the inference in a short and sharp way.

Now let's examine the passages of the *Philosophical Investigations* that are related to the argument, beginning with paragraph 243 and continuing to paragraph 258, with an important return to the issue in paragraph 298.

1. In the fragment §243, Wittgenstein introduces the question about the possibility of a language by which one person can write or express one's private sensations for oneself.

   > But is it also conceivable that there be a language in which a person could write down or give voice to his inner experiences — his feelings, moods, and so on — for his own use? — Well, can't we do so in our ordinary language? — But that is not what I mean. The words of this language are to refer to what only the speaker can know — to his immediate private sensations. So another person cannot understand the language.[2]

   Then in the next fragments Wittgenstein goes to examine the aspects of this question and reach to its answer.

---

[2]Wäre aber auch eine Sprache denkbar, in der Einer seine inneren Erlebnisse — seine Gefühle, Stimmungen, etc. — für den eigenen Gebrauch aufschreiben, oder aussprechen könnte? — Können wir denn das in unserer gewöhnlichen Sprache nicht tun? — Aber so meine ich's nicht. Die Wörter dieser Sprache sollen sich auf das beziehen, wovon nur der Sprechende wissen kann; auf seine unmittelbaren, privaten, Empfindungen. Ein Anderer kann diese Sprache also nicht verstehen(Wittgenstein, 2009, p.95)

2. In §244 dealing with the question how words refer to sensations, he regards to the manner of language learning and asserts:

> Here is one possibility: words are connected with the primitive, the natural, expressions of the sensation and used in their place. A child has hurt himself and he cries; and then adults talk to him and teach him exclamations and, later, sentences. They teach the child new pain-behaviour.
>
> "So you are saying that the word 'pain' really means crying?" —On the contrary: the verbal expression of pain replaces crying, it does not describe it.[3]

Wittgenstein introduces this assertion as "one possibility" but apparently it turns out to be a basis in the rest of argument.

The idea is that to use a word is not to *describe* a sensation nor it is to refer to a *concept* or *meaning* of a sensation but it is a mere substitution for the natural displaying of it.

3. In §246 Wittgenstein puts that speaking of "knowing" a sensation is irrelevant: human being just "has" sensations; and this means that he meanwhile wants to conclude that private language cannot be defined as about to what only I "know" it, for sensation does not belong to knowing but to having:

> In what sense are my sensations *private*? —Well, only I can know whether I am really in pain; another

---

[3]Dies ist eine Möglichkeit: Es werden Worte mit dem ursprünglichen, natürlichen, Ausdruck der Empfindung verbunden und an dessen Stelle gesetzt. Ein Kind hat sich verletzt, es schreit; und nun sprechen ihm die Erwachsenen zu und bringen ihm Ausrufe und später Sätze bei. Sie lehren das Kind ein neues Schmerzbenehmen.

"So sagst du also, daß das Wort 'Schmerz' eigentlich das Schreien bedeute?" — Im Gegenteil; der Wortausdruck des Schmerzes ersetzt das Schreien und beschreibt es nicht.(Wittgenstein, 2009, p.95)

1.2. PRIVATE LANGUAGE ARGUMENT

person can only surmise it. —In one way this is false, and in another nonsense. If we are using the word "know" as it is normally used (and how else are we to use it?), then other people very often know if I'm in pain. —Yes, but all the same, not with the certainty with which I know it myself! —It can't be said of me at all (except perhaps as a joke) that I *know* I'm in pain. What is it supposed to mean —except perhaps that I *am* in pain?

Other people cannot be said to learn of my sensations only from my behaviour —for I cannot be said to learn of them. I *have* them.[4]

That a human being can be sure of his own sensations or intents, for Wittgenstein has no peculiar significance but only concerns the manner of the uses of the words "intent", "know" and "uncertainty".

"Only you can know if you had that intention." One might tell someone this when explaining the meaning of the word "intention" to him. For then it means: *that* is how we use it.

(And here "know" means that the expression of

---

[4]Inwiefern sind nun meine Empfindungen *privat*? — Nun, nur ich kann wissen, ob ich wirklich Schmerzen habe; der Andere kann es nur vermuten. — Das ist in einer Weise falsch, in einer andern unsinnig. Wenn wir das Wort "wissen" gebrauchen, wie es normalerweise gebraucht wird (und wie sollen wir es denn gebrauchen!) dann wissen es Andre sehr häufig, wenn ich Schmerzen habe. — Ja, aber doch nicht mit der Sicherheit, mit der ich selbst es weiß! — Von mir kann man überhaupt nicht sagen (außer etwa im Spaß) ich *wisse*, daß ich Schmerzen *habe*. Was soll es denn heißen — außer etwa, daß ich Schmerzen habe?

Man kann nicht sagen, die Andern lernen meine Empfindung *nur* durch mein Benehmen, — denn von mir kann man nicht sagen, ich lernte sie. Ich *habe* sie.(Wittgenstein, 2009, p.95)

uncertainty is senseless.)⁵

Accordingly he asserts in §248 that assigning the attribute "private" to a sensation does not add something to the notion of sensation but shows its original use and meaning; or as he declares in §251 the propositions like "my mental images are private" are really grammatical not empirical.

> What does it mean when we say, "I can't imagine the opposite of this" or "What would it be like if it were otherwise?" — For example, when someone has said that my mental images are private; or that only I myself can know whether I am feeling pain; and so forth.
>
> Of course, here "I can't imagine the opposite" doesn't mean: my powers of imagination are unequal to the task. We use these words to fend off something whose form produces the illusion of being an empirical proposition, but which is really a grammatical one.⁶

Then Wittgenstein endorses the stance that there is no linguistic evidence for existence of private objects.

4. In §256 Wittgenstein continues the subject as follows:

---

[5]"Nur du kannst wissen, ob du die Absicht hattest." Das könnte man jemandem sagen, wenn man ihm die Bedeutung des Wortes "Absicht" erklärt. Es heißt dann nämlich: so gebrauchen wir es. (Und "wissen" heißt hier, daß der Ausdruck der Ungewißheit sinnlos ist.)(Wittgenstein, 2009, p.96)

[6]Was bedeutet es, wenn wir sagen: "Ich kann mir das Gegenteil davon nicht vorstellen", oder: "Wie wäre es denn, wenn's anders wäre?" a Z. B., wenn jemand gesagt hat, daß meine Vorstellungen privat seien; oder, daß nur ich selbst wissen kann, ob ich einen Schmerz empfinde; und dergleichen. "Ich kann mir das Gegenteil nicht vorstellen" heißt hier natürlich nicht: meine Vorstellungskraft reicht nicht hin. Wir wehren uns mit diesen Worten gegen etwas, was uns durch seine Form einen Erfahrungssatz vortäuscht, aber in Wirklichkeit ein grammatischer Satz ist.(Wittgenstein, 2009, p.96)

1.2. PRIVATE LANGUAGE ARGUMENT

> Now, what about the language which describes my inner experiences and which only I myself can understand? *How* do I use words to signify my sensations? — As we ordinarily do? Then are my words for sensations tied up with my natural expressions of sensation? In that case my language is not a 'private' one. Someone else might understand it as well as I. — But suppose I didn't have any natural expression of sensation, but only had sensations? And now I simply *associate* names with sensations, and use these names in descriptions.[7]

It means that if there is a natural expression for a sensation then my language is not private; and more investigations are needed to study the case in which there is no natural expression. Then he tries to examine this case in the next fragments. His account of "natural expression" here is that which has been mentioned in clause 2, namely the reaction to a sensation which is an external manifestation perceivable by others like crying, groan, grimace, etc. This case is called by Wittgenstein natural expression.

5. In §258 the he furthers the discussion; and in fact this is the core of his argument after declaring his standpoints about the relevant issues. Here he supposes the situation in which a sensation without any natural expression (i.e., considered in

---

[7]Wie ist es nun mit der Sprache, die meine innern Erlebnisse beschreibt und die nur ich selbst verstehen kann? Wie bezeichne ich meine Empfindungen mit Worten? a So wie wir's gewöhnlich tun? Sind also meine Empfindungsworte mit meinen natürlichen Empfindungsäußerungen verknüpft? a In diesem Falle ist meine Sprache nicht 'privat'. Ein Anderer könnte sie verstehen, wie ich. a Aber wie, wenn ich keine natürlichen Äußerungen der Empfindung, sondern nur die Empfindung besäße? Und nun *assoziiere* ich einfach Namen mit den Empfindungen und verwende diese Namen in einer Beschreibung.(Wittgenstein, 2009, p.98)

the state that any external reaction has not been associated to that) is to be named:

> Let's imagine the following case. I want to keep a diary about the recurrence of a certain sensation. To this end I associate it with the sign "S" and write this sign in a calendar for every day on which I have the sensation. — I first want to observe that a definition of the sign cannot be formulated. — But all the same, I can give one to myself as a kind of ostensive definition! —How? Can I point to the sensation? —Not in the ordinary sense. But I speak, or write the sign down, and at the same time I concentrate my attention on the sensation — and so, as it were, point to it inwardly. —But what is this ceremony for? For that is all it seems to be! A definition serves to lay down the meaning of a sign, doesn't it? —Well, that is done precisely by concentrating my attention; for in this way I commit to memory the connection between the sign and the sensation. —But "I commit it to memory" can only mean: this process brings it about that I remember the connection *correctly* in the future. But in the present case, I have no criterion of correctness. One would like to say: whatever is going to seem correct to me is correct. And that only means that here we can't talk about 'correct'.[8]

---

[8]Stellen wir uns diesen Fall vor. Ich will über das Wiederkehren einer gewissen Empfindung ein Tagebuch führen. Dazu assoziiere ich sie mit dem Zeichen "E" und schreibe in einem Kalender zu jedem Tag, an dem ich die Empfindung habe, dieses Zeichen. — Ich will zuerst bemerken, daß sich eine Definition des Zeichens nicht aussprechen läßt. a Aber ich kann sie doch mir selbst als eine Art hinweisende Definition geben! a Wie? kann ich auf die Empfindung zeigen? a Nicht im gewöhnlichen Sinne. Aber ich spreche, oder schreibe das Zeichen, und dabei konzentriere ich meine Aufmerksamkeit auf

1.2. PRIVATE LANGUAGE ARGUMENT

6. Thus to have a private language is taken to be nonsense; and the point is the very last statement. The conclusion here is a case of a more general principle which Wittgenstein has expressed before in §202:

   > ...to *think* one is following a rule is not to follow a rule. And that's why it's not possible to follow a rule 'privately'; otherwise, thinking one was following a rule would be the same thing as following it.[9]

   Therefore since in naming while there is no reaction or external counterpart on the basis of which other people can judge, it must be followed a rule privately, and since this is impossible, then such a naming and such a use of language is absurd.

7. In §293 which is related to language too, Wittgenstein brings an example to show the impossibility of a language —though public— about a thing that every person have access to it solely.

   > Suppose that everyone had a box with something in it which we call a "beetle". No one can ever look into anyone else's box, and everyone says he knows what a beetle is only by looking at *his* beetle.

---

die Empfindung a zeige also gleichsam im Innern auf sie. a Aber wozu diese Zeremonie? denn nur eine solche scheint es zu sein! Eine Definition dient doch dazu, die Bedeutung eines Zeichens festzulegen. a Nun, das geschieht eben durch das Konzentrieren der Aufmerksamkeit; denn dadurch präge ich mir die Verbindung des Zeichens mit der Empfindung ein. a "Ich präge sie mir ein" kann doch nur heißen: dieser Vorgang bewirkt, daß ich mich in Zukunft richtig an die Verbindung erinnere. Aber in unserm Falle habe ich ja kein Kriterium für die Richtigkeit. Man möchte hier sagen: richtig ist, was immer mir als richtig erscheinen wird. Und das heißt nur, daß hier von 'richtig' nicht geredet werden kann.(Wittgenstein, 2009, p.98)

[9]der Regel zu folgen *glauben* ist nicht: der Regel folgen. Und darum kann man nicht der Regel 'privatim' folgen, weil sonst der Regel zu folgen glauben dasselbe wäre, wie der Regel folgen.(Wittgenstein, 2009, p.88)

> —Here it would be quite possible for everyone to have something different in his box. One might even imagine such a thing constantly changing. — But what if these people's word "beetle" had a use nonetheless? —If so, it would not be as the name of a thing. The thing in the box doesn't belong to the language-game at all; not even as a *Something*: for the box might even be empty. —No, one can 'divide through' by the thing in the box; it cancels out, whatever it is.[10]

Wittgenstein finishes this fragment as follows:

> That is to say, if we construe the grammar of the expression of sensation on the model of 'object and name', the object drops out of consideration as irrelevant.[11]

This paragraph and paragraph 258 are aimed to say that it is a mistake to assume language constitution as to label objects by names, because such a labeling presupposes the existence of language; otherwise the object would have only

---

[10]Angenommen, es hätte Jeder eine Schachtel, darin wäre etwas, was wir "Käfer" nennen. Niemand kann je in die Schachtel des Andern schaun; und Jeder sagt, er wisse nur vom Anblick *seines* Käfers, was ein Käfer ist. a Da könnte es ja sein, daß Jeder ein anderes Ding in seiner Schachtel hätte. Ja, man könnte sich vorstellen, daß sich ein solches Ding fortwährend veränderte. a Aber wenn nun das Wort "Käfer" dieser Leute doch einen Gebrauch hätte? a So wäre er nicht der der Bezeichnung eines Dings. Das Ding in der Schachtel gehört überhaupt nicht zum Sprachspiel; auch nicht einmal als ein *Etwas*: denn die Schachtel könnte auch leer sein. a Nein, durch dieses Ding in der Schachtel kann 'gekürzt werden'; es hebt sich weg, was immer es ist.(Wittgenstein, 2009, p.106)

[11]Das heißt: Wenn man die Grammatik des Ausdrucks der Empfindung nach dem Muster von 'Gegenstand und Bezeichnung' konstruiert, dann fällt der Gegenstand als irrelevant aus der Betrachtung heraus.(Wittgenstein, 2009, p.107)

1.2. PRIVATE LANGUAGE ARGUMENT

an arbitrary role because there would be no criterion for its identity. Therefore, it is impossible to establish language by means of assigning signs to things.

As it can be seen, in the above represented argument there are certain accounts about language learning, the relation between word and objectivity, and rule following and personal access to mental states; and on the bases of these premises Wittgenstein arrives to the point to reject the possibility of genuine inner use of language.

Now let's reconstruct the argument in a more explicit way.

The argument consists of four premises:

α) A sole person could not follow a rule. (step 6 in the above report)

β) The word is in fact replaces in lieu of a (natural) displaying of sensations. It is not originally introduced in order to signify the meaning or the concept. (step 2)

γ) It is more likely meaningless to speak of "private"; we have no fundamentally private experience about which we are in need to have a language. (step 3)

δ) It is impossible to establish a language exclusively by means of assigning signs (words) to things and states. (steps 5 and 7)

Private language may have two senses and Wittgenstein's argument stands against both of them. The first is the language that privately used by a person considered in isolation to express his sensations (and even perceptions). The second is the language which is supposed to speak about those things which are assumed to be private. Items α, β and δ together result in impossibility of having a language in solitude; and thus the originality of inner language is rejected. The items γ and δ together results in impossibility of having a language in order to speak of private experiences.

If we accept the element β, one who is considered in isolation would have to replace his reactions to his sensations by himself, then he would ought to follow a rule in solitude which is impossible according to premise α. He would not identify his sensations as the same and all he does would be arbitrary.

Furthermore, there is no language to speak of private things, first of all because there are no such things, moreover, even if there would be private objects one could not establish a language about them because he would ought to do so by first assigning names to them—there would be no other way to speak about them because they themselves are not commonly accessible—and this leads to impossible.

There are some controversies on the interpretation of this argument.[12] However, we can now say that some interpretations (such

---

[12]For a recent and comprehensive review about the different trends in interpretation the Private Language Argument see (Ladov, 2012).

Also Law (Law, 2004) distinguishes five different forms of reading of the argument—though he finally finds the argument flawed in each of its readings. Law mentions those five different readings as follows:

1. The strongly verificationist No-Independent-Check argument

2. The weakly verificationist No-Independent-Check argument

3. Hacker's circularity argument (Hacker, 1987)

4. The stage-setting argument

5. Kenny's argument based on the unreliability of memory (Kenny, 2006, p. 194)

The first two stress the issues about rule following (item α in our reading) and the rest stress on the conditions around assigning words to things (item δ in our reading). Law observes that most of the commentators, admitting the difficulties of holding a strong verificationistic stance, support a kind of reading which Law regards under the second item of his list. According to these commentators, as Law says,

> someone can be said to follow a rule only if there exists 'a process of independent verification',(Johnston, 1993) 'usable criterion of successful performance'(Pears, 1996) or 'operational standard of

1.2. PRIVATE LANGUAGE ARGUMENT

as those presented by Hintikka (Hintikka, 2000), Kenny (Kenny, 2006), Malcolm (Malcolm, 1998), Kripke (Kripke, 1982)) emphasize what I distinguished as the element α, while some stress the element δ ((Candlish and Wrisley, 2012), (McGinn, 1997), (Budd, 1989)). Nevertheless, there are some obscurities in most of available interpretations of private language argument, in regard to which we should make a few remarks immediately after representing the argument itself. Here are these remarks:

1. Some well-known commentators ((Baker and Hacker, 1984) (Baker and Hacker, 1985), (Candlish and Wrisley, 2012)) favor to stress that Wittgenstein's argument is against the *necessarily* private language. But what difference does it make? I doubt this ascription is helpful. We should distinguish between these two notions:

    A: a language that is both necessarily originated in solitude and is necessarily used in solitude, that is can not be made public.

    B: a language that is necessarily originated in solitude no matter it then may be made public or employed privately.

    Regarding to the content of Wittgenstein's argument, it will be obvious that his target is the second one. Of course by rejecting the second case the first case would be rejected as well. The question is not about necessity or contingency, the

---

correctness'(Glock, 1996) by which the putative rule-follower's application of the rule can be checked. In short, *someone can be said to follow a rule only if it is possible (for the rule-follower, or at least for someone) to verify that they do so.*

On the other hand some commentators see the argument as speaking about the conditions of name-giving and thus consider some different passage as central. As Law points out, to find a thorough version of the so-called private language argument has itself turned out to be a confusing task.

question is that whether language can be originated in solitude. Wittgenstein answers "No", phenomenology would answer, as we are going to see, "Yes". Both of them may reject the case A, namely the language that is used necessarily in private.

By reading Wittgenstein's argument with the help of the term "necessary" it is implied that he basically intends to reject the case A. So, his stance would be more acceptable but trivial and insignificant in regard to philosophical debates. It is somewhat insignificant to say that "a language which is necessarily used in solitude is impossible" because such a claim would be irrelevant to any study of the nature of the very language that we are concerned about. It says that every language must be able to be understandable by others and must be able to be used in communication; such a claim is already granted. The main problem is about the origin, namely that whether it is possible that a language is constituted without it being actually emerged in communication. Indeed the aim of Wittgenstein is to show that language can not be *originated* in solitude, thus language is originally and essentially public.[13] So, regarding to the widely admired role of language and expression in consciousness, Wittgenstein would be able claim that being-in-community is constituting the essence of most of

---

[13]Hintikka rightly points out:

> What Wittgenstein argues is nevertheless that language is an essentially public enterprise in the sense of being publicly accessible. (Hintikka, 2000, p. 43)

However, I think the last part of the sentence is not necessary or maybe it is misleading; for a thing can be an essentially *subjective* enterprise but then happens to be publicly accessible, then being publicly accessible is not equal with being an essentially public enterprise, though follows from it. If I may I would rephrase Hintikka's remark in this way:

> What Wittgenstein argues is nevertheless is that language is an essentially public enterprise.

1.2. PRIVATE LANGUAGE ARGUMENT

human properties and that we have no self-sufficient consciousness. This is the conclusion certain well-known thinkers have developed directly or tacitly on the basis of Wittgenstein's argument. We can name here Habermas (Habermas, 1996), Apel (Apel, 1998) and Winch (Winch, 1990). It is clear that it is not the possibility of some kind of secret language that is of philosophical significance, but the possibility of "subjective language", no matter whether or not it then turns out to be public.

However, by recalling the notion of necessity the mentioned commentators want to make a point. This point, I think, is better made by means of the distinctions about the notion of private language we have already mentioned:

A': inner language, that is the language that has been originated in solitude

B': language about private objects

C': mental speech, that is the speech one performs solely and it might have been originated in public

Wittgenstein's argument obviously is not against this last case, for it is an indubitable phenomenon. It is rather against the first two cases, albeit in the sense that to reject the first case would result in rejecting the second one. Our concentration is on the first case.[14]

---

[14]Therefore, there is no need for us to enter a discussion about whether or not Wittgenstein admits private objects, mental states in particular. Thus the element γ of our reconstruction of the argument is only of a partial importance and does not situated in the center of the study. Just to mention, while most authors assert that Wittgenstein does not admit the existence of mental states, Geach (Geach, 1971, p. 3) has a different point of view:

> Of course Wittgenstein did not want to deny the obvious truth that people have a 'private' mental life, in the sense that they have for example thoughts they do not utter and pains they do not show; nor did he try to analyse away this truth in a neo-behaviouristic fashion.

The point that the mentioned authors want to make, and we are in agreement with them, is that the private language argument is not designed to reject the mental speech in the broad sens of the term. Namely, if one call the item A' above "necessarily private" and C' "contingently private", one is right to say that Wittgenstein's argument is against the "necessarily private" language. However, I do not favor this kind of calling because "necessarily private" may have a too strong meaning in the sense that to reject it would not have any philosophical significance. In a clear terminology, Wittgenstein's argument is supposed to be against subjective language as it is considered as primordial not as merely internalization of public language—and no matter this primordial, subjective language must be essentially able to become public.

2. We should notice that to rely only on the element δ which can be called, as Law(Law, 2004) suggests, Stage-Setting Argument (the argument which says in order to introduce an ostensive definition for an object we should have a background of language and grammar, then language itself could not be originated with ostensive definitions) is not sufficient to reject the inner language. As Law rightly remarks:

> That the Stage-Setting Argument ... fails the test of cogency may be demonstrated by noting that the same line of reasoning would also rule out the possibility of our starting a public language. For how did public language get started? If the only

---

... what Wittgenstein wanted to deny was not the private *reference* of psychological expressions—e.g. that "pain" stands for a kind of experience that may be quite 'private'—but the possibility of giving them a private *sense*—e.g. of giving sense to the word "pain" by just attending to one's own pain-experiences, a performance that would be private and uncheckable.

1.2. PRIVATE LANGUAGE ARGUMENT

candidate is by means of an ostensive definition, then, by the same argument, it should be impossible to start a public language. For the linguistic stage-setting required for the very first ostensive definition to succeed would also be absent in the public domain. As we clearly did manage to start a public language, the Stage-Setting Argument cannot be cogent.

The conclusion we ought to draw, of course, is that *there must be some other way* to start a language. Public language did not start with an ostensive definition. The question then arises: given that public language started not with an ostensive definition, but in some other (presumably less cerebral, more spontaneous and organic) way, *why couldn't a private language also get started in this other way?*

The authors of Stanford entry on the Private Language Argument (Candlish and Wrisley, 2012) seem to ultimately refer to a kind of stage-setting argument. They insightfully remark that:

For there to be factual assertion, there must be the distinction between truth and falsehood, between saying what is the case and saying what is not. For there to be the distinction between truth and falsehood, there must be a further distinction between the source of the meaning, and the source of the truth, of what is said.

This is true, but proofs nothing in favor of rejecting the possibility of private language. It just shows that the meaning of an object can not be exclusively grounded on the perception of it—this is in fact a significant phenomenological principle we will explain later on—and in order to constitute meaning, a more complicated process than simply pointing out to a perceived object is needed. But in order to reject subjective language stage-setting argument is not sufficient and other elements should be called up.

3. The element β of our reconstruction is highly relevant, though it is absent from the analyses of most of the commentators. The significance of Wittgenstein's argument is the effect that it is supposed to make in our understanding of the language, the very public language as a Wittgensteinian would say. The argument is supposed to show that the public language is not a generalization of the private language, but language is an issue within social behavior. Then the element β is essential. If the argument only refutes private language saying nothing about the language as such, what philosophical significance it could have?

By rejecting the possibility of private language, Wittgenstein wants to object the stance that language could be originated in solitude, then his argument wants to support a certain account of language, the account that says language is originally and necessarily social. Then its root should be sought for not in an assumedly language-less ego but in an assumedly language-less society. In the element β Wittgenstein tries to give a picture of such a language-less sociality and its transformation by adopting language. Wittgenstein's argument can be understood as if it tries to say that there is no plausible way to conceive how a language-less ego can adopt language, while there is a way to conceive this for a community.

To overlook this point will result in certain misinterpretations of the structure of the private language argument. Consider this:

> assume that it is true that a private language is possible, show that that assumption leads to certain absurdities or a contradiction, and then conclude

that it is actually false that a private language is possible. This is the way in which the argument was typically understood.(Candlish and Wrisley, 2012)

Perhaps the argument is typically understood in this way. But it is by no means corresponding to what Wittgenstein really did. There Wittgenstein intended to show that certain assumptions (which is assumed to be the only plausible assumptions) about the manner of originating a private language, *not the assumption of its existence*, lead to impossible. There is a very important difference between these two.

The argument does not primarily concern the conditions of possibility of private language, rather it concerns the conditions of originating the language privately and then the argument aims to decline the assumed explanatory power of the account that takes language as a subject to the privately origination in comparison with a proposed scenario describing the origination of language in community.

These critical remarks about the interpretation of the private language argument are to clarify the structure of argument. I did not intend here to criticize the argument itself. I just wanted to show its philosophical meaning and remove some misunderstandings about it. We will arrive to the point to deal the argument at the end of this chapter.

1.3 Husserl's Acceptance of Genuineness of Inner Language

Husserl has not dealt with the subject of inner dialogue and its probable importance in full details. However, in some occasions, that we shall mention in this section, he speaks about the language and some other related subjects more or less directly in the manner that the possibility of inner dialogue is taken for granted. Beside this, some of scholars consider this possibility as vital for phenomenology.

In this respect Derrida's viewpoint—as we mentioned in the introduction—is more notable. He equates the possibility of phenomenological reduction with the possibility of interior monologue. Derrida argues that if we rule out the possibility of inner speech and if there are good reasons to believe that language and meaning are essentially involved in outer functioning, namely in the real and empirical world, and since by epoche we suppose to suspend all empirical elements, then we will not have any *meaning* as functioning apodictically after epoche; it means that epoche will turn out to be completely useless, as it results in a total inactivity (Derrida, 1973, p. 43). Or as James Mensch says:

> For Derrida, indicative signification implies empirical existence. If expressions could not function without such signification, then the bracketing of empirical existence also brackets the functioning of expressions... Thus, the reduction would leave us with nothing at all. Its result would be a sense-less silence. To assert this, however, is to say that it is impossible. It cannot reach its intended goal. (Mensch, 2001)

About that whether Derrida exaggerates on this issue or inner speech is really so much important for phenomenology, we may judge at the end of this investigation. However, by now, we face this question: then how would the phenomenology manage to respond to Wittgenstein's argument? and if the use of inner language were so much principal why did Husserl not deal with it specifically and in detail? I think there are two possibilities; The first is that Husserl considered it unnecessary to assign an independent argument to assert the possibility of inner speech, since there were no outstanding philosophical stance refuting the genuineness of this phenomenon at his time. The second is that Husserl was confronted with a considerable number of philosophical problems in his own course

1.3. HUSSERL AND INNER LANGUAGE

and dealt with them—we are to assert, in a certain, constant and harmonic method which is called phenomenology. Nevertheless some problems have found little place to be specifically dealt with; and this is completely expectable regarding the vast area of problems Husserl was exploring.

I think that both of the mentioned possibilities may be the case; and this latter should be specially emphasized. To introduce a method is one thing and to apply it in various fields is another. The main contribution of Husserl is to elucidate the *method* of phenomenology. This includes, beside the clarification of the principles, the demonstration of the manner of exerting the method and showing the consequences in some subject-matters. However we cannot expect that the introduced method have been executed in all areas, especially when the method is supposed to be a fundamental one and thus without any priorly limited horizon. Regarding this issue, it is not odd that among the numerous issues Husserl have dealt with, such as the foundations of logic, the foundations of mathematics, consciousness of time, life-world, the essence and status of the science, intersubjective monadology and so on, certain issues more or less equally important remain less considered.

Of main importance is that we have a method and some of the basic consequences by means of which we can approach our problem, that is the possibility and the status (if any) of inner dialogue.

At first, it would be fruitful to see that whether or not the phenomenological account of language is primarily such that accept the possibility of inner dialogue and takes it as possessing some peculiarities. In this section I want to show that the stance of Husserl on this issue is in opposition to the stance of latter Wittgenstein. Then, having Wittgenstein's criticism in mind, we progress in our investigation and we have to respond to this criticism at the end of the chapter.

In this section, I mention three topics from which the possibility of inner dialogue is more or less directly inferred. Nevertheless, it is not the case that the subject of inner use of language appears for Husserl exclusively under these three topics. This is only to sketch out the starting point that should prepare us to enter the main, thematic phase of our investigation.

1.3.1 Immateriality of language

First of all, language for Husserl is a spiritual product of human. It means that it is constituted as belonged to the sphere of sprite not physic and thus its nature is not essentially dependent on external and factual conditions. Husserl in his phenomenological approach toward the nature of expression asserts this central idea:

> Language has the Objectivity as the objectivities of the so-called intellectual or cultural world, not as that of the bare physical Nature.[15]

That is to say, language is an intellectual issue and like any other intellectual issues (like art as Husserl considers it) is primarily produced in mind and then is instantiated in the external world. This viewpoint clearly refutes the viewpoints that define language in regard to its appearance in the factual society and so take inner dialogue probably as a mere reflection or imagination of what primarily and essentially is factual and social.

Just as the case for an art product, Husserl says, we should distinguish between *creation* and *reproduction*; and for a cultural entity:

> only in the form of reproduction does it have factual existence in the real world.[16](Husserl, 1969, p. 20)

---

[15]"Die Sprache hat die Objektivität der Gegenständlichkeiten der sogenannten geistigen Welt oder Kulturwelt und nicht die der blossen physischen Natur."(Husserl, 1992, p. 24)

[16]"Anderseits nur in der Form die Reproduktion hat er Dasein in der realen Welt."(Husserl, 1992, p. 25)

1.3. HUSSERL AND INNER LANGUAGE 31

Thus Husserl remarks that even when a sentence is said for the first time it is a reproduction, reproduction of what has already been produced in mind.

Husserl's analysis of language in his book Formal and Transcendental Logic (FTL hereafter) begins with the mentioned remark on the ideality of language. The discussion is based on the earlier lectures (Husserl, 2001)(1920/21) on the issue, and also Husserl came back to this topic later(Husserl, 1954). The relation of this topic, i.e., ideality of language, with the possibility of inner use of language is somehow clear; and it is not surprising that Husserl in the mentioned lecture immediately after introducing his argument on immateriality of language goes to the topic of solitary speaking:

> We distinguish between the cases where one speaks to another communicatively, and the cases where one speaks to no one, thinking in solitude, expressing oneself monologically. (Husserl, 2001, p. 12)

Nevertheless this phrase did not find its way to the latter major work FTL. It is plausible to think that Husserl found this last topic itself so complicated (or controversial[17]) that may cause some confusions in the main argument. However, regardless of the ascriptions Husserl did there about that distinction, it is obvious that Husserl believes in the originality of the distinction between communicative speaking and solitary one—the fact which is more or less entailed in the

---

[17]This is not in contradiction with what we have said before about that there were apparently no such controversy about the genuineness of the inner use of language in Husserl's age so that made him pay attention to prove it; the controversy which we mention above is about the relation between interior monologue and thinking and whether thinking itself can be equated with this monologue, or, in other words, whether thinking is linguistic from the beginning. These are some problems that Husserl briefly spoke about in the aforementioned lecture, following the above brought sentence. The complicatedness of the topics which immediately arise after the aforementioned distinction, perhaps, leaded Husserl to drop this matter while focusing on the ideality of language just in the beginning of his work FTL.

ideality of language. We will come back to this issue and will discuss it more analytically in the next section.

1.3.2 Uncommunicative expressions

It is obvious that human being can speak with itself, but the question is that whether this phenomenon is so well-established that can be taken as basis for a fundamental analysis of the nature of language or it is just a by-product of some more basic phenomena constituting the essence of language. For Husserl the phenomenon of uncommunicative use of language is so basic that he takes it for granted in investigating the language, or in other words he refers to this phenomenon as a *proof of concept* for the fundamental distinction between expression and indication:

> *Expressions* function meaningfully even in *isolated mental life, where they no longer serve to indicate anything.*[18] (Husserl, 2001b, p. 269)

Husserl develops an analysis on non-communicative speech in an unpublished manuscript from the year 1910, in which he asks:

> Kann ich nicht in meiner Aussage scharf unterscheiden den Ausdruck des Satzes selbst , z.B. des mathematischen Satzes, der physikalischen Tatsache etc., und andererseits die Funktionen der Mitteilung und des Verständnisses? (Husserl, 2005, p. 246)

> Can't I sharply distinguish in my statement between the expression of the proposition itself, e.g. the mathematical proposition, the physical fact etc., and on the other hand, the functions of proclaiming and being understood?[19]

---

[18]"Die Ausdrücke entfalten ihre bedeutungsfunction aber auch im einsamen Seelenleben, wo sie nicht mehr als Anzeichen fungieren,"(Husserl, 1913, p. 24)

[19]Translation is mine.

1.3. HUSSERL AND INNER LANGUAGE

His analysis leads to a positive answer. Accordingly, he distinguishes between the occasional character of speech and its mere expressive content. Even in the communicative expression there is a non-communicative conceptual content—which makes linguistic and conceptual communication possible. Then the non-communicative use of language, for Husserl, not only is not a marginal effect but it is a case in which the pure feature of language, *expression* of *meaning*, shows itself—though often, as Husserl states, in companion with phantasising an occasional situation.

Mohanty's explanation on this issue is very illustrative:

> In communicative speech, expressions exercise the function of pronouncement in addition to their meaning-function. (But we have emphasized that even in communicative speech it is the meaning-function that predominates in the case of predicative statements.) Are the two functions, the meaning and the pronouncing-functions, inseparable? Or, is it possible to find the one even in the absence of the other? Husserl's answer seems to be that although it is possible to find the meaning-function in the absence of the pronouncing-function, it is not possible to find the pronouncing-function in the absence of the meaning-function.(Mohanty, 1976, p. 13)

1.3.3 Prayer, inwardness and self-directing

There are some remarks in Husserl's unpublished writings in which he speaks of inner-direction in the genuine prayer. Here is his explicit phrase:

> In truly intimate prayer, i.e. in genuine one, I pray not outwardly, rather inwardly directed.[20]

---

[20]"Im wirklich innigen Gebet, dem echten, ist das betende Ich nicht nach aussen, sondern nach innen gerichtet."(Husserl, 2014)

It means that we have a specific kind of uncommunicative speech in which we face God not as a worldly real being; and thus this kind of activity is by no means directed to the outer real world and can be, or must be, genuinely done in solitude. Of crucial importance is that this case is not for Husserl a secondary, non-original phenomenon. If it was so one could say that praying is just a repetition of what one have heard in community and there is nothing primary in it. However at the end of the paragraph from which the above quotation has been brought Husserl says:

> This inner-direction is parallel with the phenomenological inner-direction, through which my interior way goes to all Others (as inner-other, not as external human being or as spatio-temporal real being) and thus into the world and into the strange human beings.[21]

It can be seen that inner-direction, for Husserl, is such a genuine feature that is taken as primordial datum for further investigation. We will deal with the issue of inner alterity and inner-direction, in so far as it relates to inner dialogue, in 1.6. By now we can say that it is clear that for Husserl inner dialogue is possible and also of principal value. Now let's see what is the phenomenological account of language that allows the possibility of inner use and what would be a phenomenological response to Wittgenstein's argument against the originality of inner language.

---

[21]"Diese Innenrichtung ist parallel mit der phänomenologischen Innenrichtung, bei welcher durch mein Innen hindurch der Weg geht in alle Anderen (als Innen-Andere, nicht als äußerliche Menschen, als raumzeitlich reale) und dadurch erst auf die Welt und auf eigenes und fremdes Menschendasein."(Husserl, 2014)

1.4 The Possibility of Inner Use of Language: Descriptive Phenomenology of Expression

In the following I will apply some of prominent ideas of phenomenology in order to clarify the basic issues belonging to language and investigate the primordial elements of dialogue so that we will be able to evaluate the possibility of inner dialogue and elucidate its probable function.

1.4.1 Language and the distinction between meaning and sense

In this section we want to show that there are significant ambiguities in the use of the word "meaning"; and before any investigation in order to reach a theory on meaning we should determine our primary account of the concept and rule out the ambiguities around the word. These ambiguities will be recognized specially when we consider the cases in which an expression is said to be "meaningless". We usually apply the attribute "meaningless" in different situations in the manner that what is taken as meaningless in a certain respect is considered as meaningful in another respect; as if we speak of more than one type of meaningfulness and thus meaning.

In the following I will carry out an analysis on this matter and, in this respect, will refer to Husserl's ideas on the issue.

It is not so uncommon that in ordinary speaking we assign the attribute "meaningless" to a notion or judgment which is *impossible* or *absurd*. However, at the same time we acknowledge that, in some sense, such expressions are more "meaningful" than those expressions which seem to be arbitrary formed. Consider for example: "An injured circle which its hooves are sleepy". Moreover, in such cases this is the meaningfulness of the statement which serves as a ground to depart to grasp its impossibility.

Let's compare these two sentences:

1. π is an algebraic number.

2. Some rational numbers are sorrowful.

The former sentence, though necessarily false, seems to be more contentful than the latter. We can say that the former is false for we accept its being meaningful. However, for the latter one may say that it is neither true nor false for it is meaningless. But yet it has its conceptual content, and in some sense it does have meaning, though according to the structure of its meaning we judge it as nonsense.

The question is what is taken to be meaning of a statement and what statements are *meaningless*, or what is the original function of wording and in what sense a wording is *meaningful*. Our descriptive analysis goes as follows.

1. In the first glance, we take as meaningful those expressions which refer to a determined object or state of affairs. One may tempt to say that the meaning of an expression is its objective reference.

2. In another level, those notions or judgments that are related to existing or even possible objects or states of affairs are taken as meaningful; and the expressions pertained to impossible phenomena are ruled out as meaningless. In this regard "a triangle with two equal side" is a meaningful expression, but "square circle" is meaningless.

3. In another sense, meaning is also attributed to those locutions which express some, though absurd, contents. In other words, the contents of such expressions seem to be, roughly speaking, conceptually well-established, as if they have been purposefully

1.4. DESCRIPTIVE PHENOMENOLOGY OF EXPRESSION

formed, instead of being an irrelevant set of components that, although grammatically valid, does not seem to have any significant purpose behind it. We have already brought two examples of such expressions that in this sense are said to be meaningless: "An injured circle which its hooves are sleepy" and "some rational numbers are sorrowful". In this regard the expression "square circle" is meaningful, it genuinely belongs to the context of geometry and in that context, and by considering the rules of that very context, such an expression is recognized as impossible or absurd.

4. In another level, meaning also includes those expressions which are not formed in a well-established manner, but still are grammatically valid wordings. Compare our previous example "some rational numbers are sorrowful" with a locution like "are some". In the former case one would like to say that the meaning there has been mistakenly formed, but there is still a kind of meaning. Such expressions seem to be results of some category mistakes. They can not be said to be true or false, rather they are *nonsense*. However there is an account of meaning which also includes nonsense statements. The account which asserts that mistakenly formed meanings are still meanings, and, as it happens, to recognize their being *inauthentic* is not as self-evident as for "are some"; it needs some knowledge and investigation on the background and context of the expressions.

An expression, which seems to be nonsense at first glance, may be notably contentful in a certain context. For example the sentence "some rational numbers are sorrowful" which is nonsense from the scientific point of view may have a meaning in some kind of numerology. Therefore it is not acceptable to neglect such an expression from the beginning as a result of accounting it as meaningless. Nevertheless, this does not restrict our right, considering the constitutional origin of an expression, to assert that that expression is nonsense in a given context. To say that every valid expression is meaningful is not equal to say that every expression is senseful.

However, in this last account every valid expression, whether or not confer a well-established content, have meaning. Only those expressions have not meaning which are not genuinely expressions.

As we see there are four accounts of meaning each of which is more general than the previous one. At the beginning of the investigation it is appropriate to take "meaning" in its widest sense, and distinguish it from other concepts by choosing other names for them. Note that this does not mean that "meaning" is a genus that other mentioned concepts are types of that. The distinction here is primarily a differentiation in applications of the term. These four concepts may be originally different concepts which are attributed to expressions. However this latter concept to which we assign the word "meaning" is more general in the sense that every expression that falls under one of the previous concepts falls under the latter one too, but this is not always the case in the opposite way. Anyway, the relation between these concepts as such is a peculiar issue which needs to be investigated.

In the first investigation of his *Logical Investigations*, Husserl tries to eliminate the confusions around the "meaning" and attempts to refine our concept of it. Before entering to a specific study on expression and meaning, in the introduction of the second volume of his book, He has distinguished between *meaning-intention*[22] and *meaning-fulfillment*[23] (Husserl, 2001b, p 250) of an expression. To clarify the difference between these two features and the difference between these two from one side and the object itself from the other side is the main theme of the first investigation. Husserl,

---

[22]Bedeutungsintention,
[23]Bedeutungserfüllung,

1.4. DESCRIPTIVE PHENOMENOLOGY OF EXPRESSION

having indicated "the ambiguities in talk about what an expression expresses", says:

> The application of the terms 'meaning' and 'sense', not merely to the content of the meaning-intention inseparable from the expression, but also to the content of the meaning-fulfillment, engenders a most unwelcome ambiguity. It is clear from previous indications, where we dealt with the fact of fulfillment, that the acts on either side, in which intending and fulfilling sense are constituted, need not be the same. What tempts us to transfer the same terms from intention to fulfillment, is the peculiar way in which the unity of fulfillment is a unity of identification or coincidence.[24] (Husserl, 2001b, p. 291)

In the explanation of this phrase, we should say that because in most of expressions there is a coincidence between intention and fulfillment—a coincidence which makes the expression to be considered as referring to a real object or states of affairs—the difference between intending sense of the expression and fulfilling sense of it is usually overlooked. For example when we say "the sky is blue", this primarily expresses an intention, though we can express a perceptional experience of ours exactly with the same locution. We can say this sentence and the audience properly understand us without having any perception or intuition. Therefore to reveal a sensation or experience is only one of the functions of the expression,

---

[24]"Die Anwendung der Termini Bedeutung und Sinn nicht bloß auf den Inhalt der Bedeutungsintention (die vom Ausdruck als solchem unabtrennbar ist), sondern auch auf den Inhalt der Bedeutungserfüllung ergiebt freilich eine sehr unliebsame Aequivocation. Denn, wie schon aus den vorläufigen Andeutungen hervorgeht, die wie der Erfüllungstatsache widmeten, sind die beiderseitigen Acte, in welchen sich intendirender und erfüllende Sinn constituiren, keineswegs dieselben. Was aber zur Uebertragung der selben Termini von der Intention auf die Erfüllung geradezu hindrängt, ist die eigenart der Erfüllungseinheit, als Einheit der Identificirung oder Deckung."(Husserl, 1913, p. 52)

though most of expressions in ordinary speaking are of this type. In any case we should note that

> Expressions and their meaning-intentions do not take their measure, in contexts of thought and knowledge, from mere intuition—I mean phenomena of external or internal sensibility—but from the varying intellectual forms through which intuited objects first become intelligibly determined, mutually related objects. And so expressions, even when they function outside of knowledge, must, as symbolic intentions, point to categorially formed unities.[25](Husserl, 2001b, p. 289)

Accordingly, we should separate meaning in the sense relating to real or possible phenomena from meaning in its widest sense. We can assign the word "sense" for the former case and "meaning" for the latter. These two notions fit in, respectively, with the 2nd and the 4th accounts which we have listed above. The 1st account of our list speaks of "reference". We can reserve the word "significance" to indicate the 3rd account of our list.

So, in order to refute the ambiguity around what may be called by the term "meaning" we distinguish between these notions:

1. Reference

2. Sense

3. Significance

---

[25]"Die Ausdrücke und ihre bedeutungintentionen messen sich, wie wir hören werden, im Denk- und Erkenntnis-zusammenhange nicht blofs den Anschauungen (ich meine den Erscheinungen der äusseren und ineren Sinnlichkeit) an, sondern auch den verscheidenen intellectiven Formen, druch welche die bloss angeschauten Objecte allerest zu verstandesmässig bestimmten und aufeinander bezogenen Objecten werden. Und demgemäss deuten die Ausdrücke, wo sie ausserhalb der Erkenntnisfunction stehen, auch als symbolische Intentionen auf die Kategorial geformten Einheiten Hin."(Husserl, 1913, p. 49)

1.4. DESCRIPTIVE PHENOMENOLOGY OF EXPRESSION

4. Meaning

About choosing the terms, it may seem that the word meaning is more appropriate for the 3rd concept, for how could we call an expression meaningful when the speaker seems not to *mean* something by it? The word meaning recalls the condition of being purposefully formed which we mention as the characteristic of the 3rd concept in our list. However, because of some strong reasons which will be clear during the investigation, and also, more importantly, to be in accordance with Husserl's terminology, we assign the word "meaning" to the latter concept, in the sense that:

> It is part of the notion of an expression to have a meaning: this precisely differentiates an expression from the other signs mentioned above. A meaningless expression is, therefore, properly speaking, no expression at all: it is at best something that claims or seems to be an expression, though, more closely considered, it is not one at all. Here belong articulate, word-like sound-patterns such as 'Abracadabra', and also combinations of genuine expressions to which no unified meaning corresponds, though their outer form seems to pretend to such a meaning, e.g. 'Green is or'.[26] (Husserl, 2001b, p. 292)

Therefore, we should take a comprehensive account of the meaning from the outset and does not impose some preconditions to the

---

[26]"Zum Begriff des Ausdrucks gehört es, eine Bedeutung zu haben. Eben dies unterscheidet ihn ja von den sonstigen Zeichen, wie wir oben ausgeführt haben. Ein bedeutungsloser Ausdruck ist also, eigentlich zu reden, überhaupt kein Ausdruck; bestenfalls ist er ein Irgendetwas, das den Anspruch oder Anschein erweckt, ein Ausdruck zu sein, während es dies, näher besehen, gar nicht ist. Hierher gehören wortartig klingende artikulierte Lautgebilde, wie *Abracadabra*, andererseits aber auch Komplexionen wirklicher Ausdrücke, denen keine einheitliche Bedeutung entspricht, während sie eine solche, bei der Art, wie sie sich äußerlich geben, doch zu prätendieren scheinen. Z. B. Grün ist oder." (Husserl, 1913, p. 54)

meaningfulness. However, as regards what is *not* meaningful, we should notice that we have a meaning at hand only in the presence of a genuine *expression*. That is, a synthetic well formed *object* by itself need not to be meaningful because it is possible that it be only a result of a mathematical function without being possible to be truly expressed. Sundholm (Sundholm, 2002) has a a nice discussion in this respect with some illustrative examples.

Husserl does not set forth an explicit distinction in order to distinguish what I called significance, but such a distinction is quite compatible with his analyses and also it is indispensable in order to avoid some confusions. Although the distinction between the significance and the meaning has not been explicitly tematized by Husserl, it is discussed in some places in his works. What I call significance is similar to what Beyer (Beyer, 2015) calls *respective meaning*. Most importantly in the case of indexical expressions, besides the "general meaning function" (*die allgemeine Bedeutungsfunktion*) (Husserl, 1913, p. 88), we have a "respective meaning" that is at work. However even in the case of indexicals there is an ideal meaning which makes the utterance understandable but does not cover the all intentional character of the expression. For example when I say "this wall is white", the sentence is meaningful, but in order it to be properly significant and thus judgable it should be uttered in a determinate occasion. In any case, the significance or the respective meaning can not substitute the general meaning and should not be confused with it. The context-sensitive or occasion-bounded significance of an utterance is itself capable to function partly due to the ideal meaning of the expression.

Related to this topic is Husserl's discussion about the *empirisehe Bedeutung* (Husserl, 1986, pp. 202–219). He there says:

> Die empirische Bedeutung ist keine Idee im Sinne eines Eidos. Sie kann nur aus einem unmodifiziert setzenden Akt entnommen werden. (Husserl, 1986, p. 211)

> The empirical meaning is not an idea in the sense of

1.4. DESCRIPTIVE PHENOMENOLOGY OF EXPRESSION

an eidos. It can only be taken from an unmodified, positional act.

In our terminology it means that the significance of an expression is not in general ideal (as opposed to its meaning); and an expression would have a significance only if it is connected to a positional, i.e. non-modified, act. Whereas "meaning" is independent of the positionality of the act. The themes "positionality" and "modification" will be discussed in the rest of the present work (including sec. 1.6 2.2.1 2.2.3).

However, if we can eliminate the confusion between *meaning*[27] and *sense*[28], it will be a great success, considering the prevalence of the meaning theories which, at their starting point, take meaning as related to the existing, or assumable as existing, object or state of affairs.[29] As we have seen, nevertheless, this is a narrow account of meaning and as far as related to the problem of *expression* it is a flawed and misguiding account. To fix a clear account of meaning and keep it apart from any denotational or verificationist conception was one of Husserl's main concerns, from his early work to his later more developed studies. Here we give examples from 1891 and from some forty years latter.

---

[27]Bedeutung,

[28]Sinn,

[29]For a distinguished example of such theories we should mention *picture theory of meaning*—and its various forms, such as truth conditional theory of meaning— as it has been formulated in Wittgenstein's *Tractatus* and some other works in early contributions to analytic philosophy.

The proposition 3.12 of Tractatus asserts:

> Das Zeichen, durch welches wir den Gedanken ausdrücken, nenne ich das Satzzeichen. Und der Satz ist das Satzzeichen in seiner projektiven Beziehung zur Welt.

> I call the sign with which we express a thought a propositional sign.—And a proposition is a propositional sign in its projective relation to the world.Wittgenstein (2015)

In his review to Schröder's book on algebraic logic, Husserl also criticizes him for having a denotational conception of meaning, and brings a passage which is almost similar to that appeared a ten years later in *LI* (the piece we just quoted). He there says:

> There is further connected with this unclarity about the concept of signification the fact that Schröder classifies names such as "round square" as "meaningless" ("unsinnige"), in comparison with names with one or more senses. Obviously he here is confusing two very different things: namely (1) whether a signification (a "sense") belongs to a name, and (2) whether or not there exists an object corresponding to a name. "Meaningless" names in the strict sense are names without a signification - pseudo names, such as "Abracadabra." But "round square" is a univocal common noun to which, however, nothing can in truth correspond.(Husserl, 1994a, p. 60)

In *Experience and Judgment*, while speaking of ideality of meanings, including propositions as distinguished from the judgment, he explains the omnitemporality of meanings which has nothing to do with the truth possibly attributed to them or the judgment may be performed on the basis of them:

> it should be noted that this omnitemporality does not simply include within itself the omnitemporality of validity. We do not speak here of validity, of truth, but merely of objectivities of the understanding as suppositions and as possible, ideal-identical, intentional poles, which can be "realized" anew at any time in individual acts of judgment— precisely as suppositions; whether they are realized in the self-evidence of truth is another question. A judgment which was once true can cease to be true, like the proposition "The automobile is the fastest means of travel," which lost its validity in the age

1.4. DESCRIPTIVE PHENOMENOLOGY OF EXPRESSION

of the airplane. Nevertheless, it can be constituted anew at any time as one and identical by any individual in the self-evidence of distinctness; and, as a supposition, it has its supertemporal, irreal identity.

...

In order to apprehend the proposition 2 < 3 as a proposition which, perhaps, we wish to divide according to its grammatical sense, we do not have to deal comparatively with the acts of judgment which judge that 2 < 3;[30] (Husserl, 1973, pp. 261-2)

The fact that meaning as such differs from sense, and that expressions without any intuition behind them can be meaningful, does not result in that the meaning is located in the word, or that signs themselves are expressive or meaningful. Husserl declares:

Our conception will perhaps be censured for its extreme nominalism, for identifying word and thought. To many it will seem quite absurd that a symbol, a word, a

---

[30]Es ist dabei zu beachten, daß diese Allzeitlichkeit nicht ohne weiteres in sich schließt Allzeitlichkeit der Geltung. Von der Geltung, der Wahrheit sprechen wir hier nicht, sondern bloß von den Verstandesgegenständlichkeiten als Vermeintheiten und möglichen ideal-identischen intentionalen Polen, die als dieselben allzeit wieder "realisiert" werden können in individuellen Urteilsakten — eben als Vermeintheiten; ob realisiert in der Evidenz der Wahrheit, ist eine andere Frage. Ein Urteil, das einmal wahr gewesen ist, kann aufhören wahr zu sein, wie etwa der Satz "das Auto ist das schnellste Verkehrsmittel" im Zeitalter der Flugzeuge seine Gültigkeit verliert. Gleichwohl kann er als dieser eine, identische von beliebigen Individuen allzeit in der Evidenz der Deutlichkeit wieder gebildet werden und hat als Vermeintheit seine überzeitliche, irreale Identität.

...

Ganz anders, wo es gilt, den Sinn einer Aussage herauszufassen und zum Gegenstand zu machen. Um den Satz 2 < 3 zu erfassen als diesen Satz, den wir etwa nach dem grammatischen Sinn zergliedern wollen, haben wir nicht Urteilsakte, die urteilen, es sei 2 < 3, vergleichend zu behandeln; (Husserl, 1939, pp. 313-5)

sentence, a formula should be understood, while in our view nothing intuitive is present beyond the mindless sensible body of thought, the sensible stroke on paper etc. But we are far from identifying words and thoughts, as our statement in the previous chapter show. We do not at all think that, where symbols are understood without the aid of accompanying images, the mere symbol alone is present: we think rather that an understanding, a peculiar act-experience relating to the expression, is present, that it shines through the expression, that it lends it meaning and thereby a relation to objects.[31] (Husserl, 2001b, p. 302)

Or as he reemphasizes:

[Although] it should be quite clear that over most of the range both of ordinary, relaxed thought and the strict thought of science, illustrative imagery plays a small part or no part at all, and that we may, in the fullest sense, judge, reason, reflect upon and refute positions, without recourse more than symbolic presentations. This situation is quite inadequately described if one talks of the 'surrogative function of signs', as if the signs themselves did duty for something and as if our interest

---

[31]Man wird unserer Auffassung vielleicht den Vorwurf des extremen Nominalismus machen, als ob sie Wort und Gedanken identifiziere. Manchem wird es geradezu als absurd erscheinen, daß ein Symbol, ein Wort, ein Satz, eine Formel verstanden sein soll, , während nach unserer Lehre anschaulich nichts anderes da. sei, als der geistlose sinnliche Körper des Gedankens, dieser sinnliche Zug auf dem Papier u. dgl. Indessen sind wir, dies bezeugen die Ausführungen des vorigen Kapitels, weit davon entfernt, Wort und Gedanken zu identifizieren. Keineswegs ist für uns in den Fällen, wo wir Symbole ohne Stütze begleitender Phantasiebilder verstehen, das bloße Symbol da; vielmehr ist das Verständnis da, dieses eigentümliche, auf den Ausdruck bezogene, ihn durchleuchtende, ihm Bedeutung und damit gegenständliche Beziehung verleihende Akterlebnis. (Husserl, 1913, p. 65)

1.4. DESCRIPTIVE PHENOMENOLOGY OF EXPRESSION

in symbolic thinking were directed signs themselves. Signs are in fact not objects of our thought at all, even surrogatively; we rather live entirely in the consciousness of meaning, of understanding, which does not lapse when accompanying imagery does so. [32] (Husserl, 2001b, p. 304)

To clarify this very *consciousness of meaning*, this *peculiar act-experience relating to the expression*, is the main objective of this project and it will be, we hope, more explained, more we progress in the investigation. But about the primary distinction between expression, which is in close relation to meaning, and sign in general, which should not be taken as the bearer of meaning, we will argue in the next part.

For now, let's explain a bit more about the distinction between *meaning* and *sense*[33] which we should keep in mind while investigating about the origin of meaning. The first is that notion that we have in mind when we express our intended content regardless that whether it is fulfilled, or even fulfillable, or not. For instance the meaning of "gold mountain" or "the horse is flying". The second notion is that which we have in mind when we speak about a percep-

---

[32]Man muß sich durchaus klar machen, daß in weitesten Strecken nicht bloß lässigen und alltäglichen, sondern streng wissenschaftlichen Denkens die veranschaulichende Bildlichkeit eine geringe oder schlechterdings keine Rolle spielt, und daß wir im aktuellsten Sinn urteilen, schließen, überlegen und widerlegen können auf Grund von "bloß symbolischen" Vorstellungen. Es ist eine sehr unangemessene Beschreibung dieser Sachlage, wenn man hier von einer stellvertretenden Funktion der Zeichen gesprochen hat, als ob die Zeichen selbst für irgendetwas surrogierten, und das Denkinteresse im symbolischen Denken den Zeichen selbst zugewendet wäre. In Wahrheit sind diese aber in keiner und auch nicht in stellvertretender Weise die Gegenstände der denkenden Betrachtung, vielmehr leben wir ganz und gar in dem bei allem Mangel an begleitender Anschauung nicht fehlenden Bedeutungs- bzw. Verständnisbewußtsein. (Husserl, 1913, p. 68)

[33]*Bedeutung* and *Sinn*; and this is quite irrelevant to the Fregean distinction which uses the same words.

tion, a fulfilled intention, a "sense" in proper use of the word. This latter notion is not meaning while meaning of an expression is not necessarily dependent on the actual perception or intuition of its content. Only if you understand the meaning of a proposition you can conceive it or judge it as true or false. It is so about a concept, for if a concept is meaningless to you, you can not conceive it or have a sense of it. But not vice versa: it is possible that you take a concept or judgment as meaningful but without having the sense of it and it maybe even impossible and thus unconceivable. Like our examples above: "a triangle with a side greater than the sum of two other sides", "π is an algebraic number" or "some rational numbers are sorrowful". These expressions are meaningful but not senseful or may be even not of significant meaning (for the last example).

Therefor we should be careful, when we attempt to clarify the origin of meaning and its function in dialogue or its role in the alleged inner use of language, do not confuse these issues with the origin and status of *sense* or *significance*; these two latter notions come into expression thanks to meaning and as far as we investigate on expression qua expression the priority is of meaning.

In fact, we should consider the meaning as primitive in comparison to other functions that an expressive locution may have. Accordingly, the meaning is connected to the transcendental intentionality, namely intention in the deepest level regardless to the other factors that may be attached to an intention: no matter that the intention is fulfilled in this or that way (in which besides meaning we would also have a *sense*), no matter that the intention is based on some situational grounds (in which we would also have a *significance*) and no matter whether the intention is toward an object which is constituted as a real object (in which we would also have a *reference*).

Based on its meaningfulness, and rarely independent of that, a locution may have different functions. Nevertheless, these functions, though very interesting in their own right, are not the direct subject

1.4. DESCRIPTIVE PHENOMENOLOGY OF EXPRESSION

of our investigation as far as the theory of meaning is concerned. A brilliant study of the various functions of speech has been done by Karl Bühler. His developed analyses are presented in *Theory of Language: The representational function of language* in which he distinguishes between three functions of speech: *profession* (*Kundgabe*), *triggering* (*Auslösung*) and *representation* (*Darstellung*),[34] (Bühler, 2011, p. 35) and the sign in respect to each of these functions, he remarkably proposes, should be called *Symptom*, *Signal* and *Symbol*. Husserl was aware of Bühler's ideas, at least as represented in Bühler's review of Marty' book (1909), and he referred to these ideas in one of his manuscripts written in 1910 (Husserl, 2005, pp. 241–285). It can be seen that although he has some critics, Husserl admires Bühler's work. Another thinker who studied the pragmatic aspects of signs in an elaborated way, as it is very well-known, is Peirce. Husserl was not aware of Perice's those works—Husserl knew Peirce mainly through the works of Schröder thus only as a (algebraistic) logician. However it can be demonstrated that Peirce's work are compatible with the phenomenological theory of meaning. What is important is to notice that Peirce's work, as well as that of Bühler, mainly are concerned with what I have distinguishes as the "significance" of the expression (and what can be included in it such as sense and reference), but Husserl argues that yet meaning is different and it is more fundamental in respect with the transcendental constitution. Studies about the functions of speech act are very important and Husserl was undoubtedly aware of their importance and he himself made some contributions in that field. However, he had made it explicit that such studies are not of the primary interest while investigating the meaning theory.

In a lecture on the theory of meaning Husserl says:

> Da uns Bedeutung und Bedeuten interessiert und nur

---

[34] In the mentioned book the author says that he prefers the following terms: expression (Ausdruck), appeal (Appell) and representation. But I preferred to use the old terms in the text in order to avoid the possible confusions.

> interessieren soll, so kommen für uns die Ausdrücke in ihren Funktionen der Kundgabe und Kundnahme, wie ich schon sagte, nicht weiter in Betracht. (Husserl, 1986, p. 11)
>
> Since our interest is in the meaning and what is meant, and should be only in that, the functions of the expression as proclamation and calling are no more the subject of our consideration.

Then in our descriptive study we should pay attention not to confuse meaning, and expression as such, with its various use; and accordingly when studying the constitutional root of expression we are not allowed to appeal to the explanations commencing with some founded functions, rather we should seek for the explanation in the level of transcendental intentionality—if the place of expression is so much deep, which according to the transcendental phenomenology it is.

1.4.2 Expression and indication

Just at the beginning of the first investigation in his logical investigations, Husserl declares one should not confuse between the notions of sign and expression when he intends to investigate the nature of expression and meaning. This is a central thesis that remains important in the whole process of investigation of meaning, language and logic.

Firstly we should consider some clarifications about the terms:

> The terms 'expression' and 'sign' are often treated as synonyms, but it will not be amiss to point out that they do not always coincide in application in common usage. Every sign is a sign for something, but not every sign has 'meaning', a 'sense' that sign 'expresses'.[35] (Husserl, 2001b, p. 269)

---

[35] Die Termini Ausdruck und Zeichen werden nicht selten wie gleichbedeutende

1.4. DESCRIPTIVE PHENOMENOLOGY OF EXPRESSION51

Regarding their intentional content or the genuine application for which we use them, expressions are not essentially involved with signs and on the other hand signs can refer to a variety of events which does not essentially contain meaning or expression. For example thunder can be considered as a sign of the probability of raining but it does not *express* it or *mean* it. On the other hand, one can mean something and express it, while for this reason one can choose an option out of a number of options from signs and does this in regard to social and cultural contexts or individual tastes or aesthetic purposes and so on. A sign can be posited, e.g. a flag for a nation, or inferred, e.g. a symptom in medicine, but it does not express something in genuine sense. Just like this, crying can be taken as a sign for pain but it is obviously different from to express to have pain.

If we disregard the distinction between meaning and sense—which we have explained in the previous section—and take meaning as identical with sense, it would be plausible to consider expression as indicator for something i.e. some perception, some intuition or some sense in general; but since expression is supposed to originally be in relation with meaning, considering the peculiarities of meaning, we cannot take expression as simply being signs for what are beforehand existing outside the realm of the expression and meaning.

That in ordinary language expression and the system of signs come together should not misguide us to the thought that those are identical. In fact most of misunderstandings in the philosophy of language and also in meaning theory arise from the confusion between expression and indication. We shall come back to this issue latter. Husserl writes:

> To mean is *not a particular way of being a sign in the*

---

behandelt. Es ist aber nicht unnütz zu beachten, daß sie sich in allgemein üblicher Rede keineswegs überall decken. Jedes Zeichen ist Zeichen für etwas, aber nicht jedes hat eine "Bedeutung", einen "Sinn", der mit dem Zeichen "ausgedrückt" ist. (Husserl, 1913, p. 23)]

> *sense of indicating something.* It has a narrower application only because meaning—in communicative speech—is always bound up with such an indicative relation, and this is in its turn leads to a wider concept, since meaning is also capable for occurring without such a connection.[36]
> (Husserl, 2001b, p. 269)

Therefore Husserl differentiates between expression and indication and thus distinguishes expression from an external instrument for it, namely indication which operates by means of the already posited signs. Indication has some parameters which should not be taken as primarily belonging to expression.

Accordingly Husserl asserts:

> *Expressions* function meaningfully even in *isolated mental life, where they no longer serve to indicate anything.*[37]
> (Husserl, 2001b, p. 269)

The sentence brought above is the clue to understand Husserl's characteristic account of expression. Also the section §8 of the mentioned investigation should be interpreted due to this clue. In that section Husserl asserts that there is no inner expression which operates by means of the words. If one do not note this principal condition, the whole argument of Husserl, which so sounds like an argument against private language, will be misunderstood. This is Husserl's words:

> Shall we say that even in solitary mental life, one still uses expressions to intimate something, though not to

---

[36] Das Bedeuten ist nicht eine Art des Zeichenseins im Sinne der Anzeige. Nur dadurch ist sein Umfang ein engerer, daß das Bedeuten — in mitteilender Rede allzeit mit einem Verhältnis jenes Anzeichenseins verflochten ist, und dieses wiederum begründet dadurch einen weiteren Begriff, d es eben auch ohne solche Verflechtung auftreten kann. (Husserl, 1913, p. 23)

[37] Die Ausdrücke entfalten ihre Bedeutungsfunktion aber auch im einsamen Seelenleben, wo sie nicht mehr als Anzeichen fungieren. (Husserl, 1913, p. 24)

1.4. DESCRIPTIVE PHENOMENOLOGY OF EXPRESSION

a second person? Shall one say that in soliloquy one speaks to oneself, and employs words as signs, i.e. as indications, of one's own inner experience? I cannot think such a view acceptable.[38] (Husserl, 2001b, p. 279)

This does not entail that to speak to oneself is senseless, but that it would be senseless if we assume that there is essentially a principal role for words. This is directly related to the idea of distinction between expression and indication: expression, being essentially independent from indicative system, "also plays a great part in uncommunicative, interior mental life" (Husserl, 2001b, p. 278) but indication depends, at least on its origin, on communicative interactions.

Therefore, we have a kind of inner speech which consists of expressing to oneself but without the mediate role of signs. Husserl declares that words can not genuinely operate as indicators in the sphere of mind.

Nevertheless as a matter of fact we can say something in our minds. Does this violate Husserl's viewpoint? I think it does not; and there is a plain way to solve this seeming conflict:

We have the imaginations of words and sentences. This is what Husserl acknowledges and there is nothing in his system that be in contradiction with this. When I say something in my mind, it is—in some occasions—a merely reproduction of what I already expressed to myself. It is not the case for all inner sentences because they maybe reproduction of what I have already heard or a mere combination of what have already came to my mind. When I say a sentence in my mind, it is not by that sentence that I express the meant content to myself. The act of expression is not done by that

---

[38] Oder sollen wir -etwa sagen, daß wir auch im einsamen Seelenleben mit dem Ausdruck etwas kundgeben, nur daß wir es nicht einem Zweiten gegenüber tun? Sollen wirsagen, der einsam Sprechende spreche zu sich selbst, es dienten auch ihm die Worte als Zeichen, nämlich als Anzeichen seiner eigenen psychischen Erlebnisse? Ich glaube nicht, daß eine solche Auffassung zu vertreten wäre. (Husserl, 1913, p. 35)

sentence. That sentence maybe a consequent of an expression, of course, a consequent by choice not necessarily. It is the consequent of that we imagine ourselves in the position of saying this content to others, then words operate for this imagination. Husserl says that we should avoid the belief that *imagined* words play a roll in inner expression. Then Husserl draws a line between inner expression which is wordless and imagined words or speech which can follow an inner expression.

Peter Geach in his book *Mental Acts* (Geach, 1971) says:

> What is more important is the difference between speech and thought as regards temporal duration. Spoken words last so long in physical time, and one word comes out after another; the time they take is, as Aquinas would say, the sort of time that is the measure of local motion—one could sensibly say that the utterance of the words was simultaneous with the movement of a body, e.g. the hand of a clock, from one place to another. The same would go for the duration of mental images of words, or any other mental images; one could sensibly say "I rehearsed the words in my head as I watched the beetle crawl from one side of the table to the other".
>
> With a thought it is quite different.... we could hardly suppose that in a thought the Ideas occur successively, as the words do in a sentence; it seems reasonable to say that unless the whole complex content is grasped all together—unless the Ideas, if Ideas there are, are all simultaneously present—the thought or judgment just does not exist at all.

As we see we find Geach's argument here in agreement with our assertion that words play no essential role in thinking and judgment. We are sympathetic with Geach and acknowledge his criticism to the extent that his criticism is against attributing a grammar analogous

1.4. DESCRIPTIVE PHENOMENOLOGY OF EXPRESSION

with the grammar of spoken language to the thought. However this should not lead us to deny pure grammar of thought if we mean by this latter a peculiar structure that should be studied in regard to its own specialties. To attribute the grammar of spoken language to the thought is wrong, for such a grammar is basically related to structure of signs and in this respect is a branch of *semiotics* (see next section) whereas the pure grammar is concerned with "meaning' categories'. However we are still blameless in applying the word language for certain acts of thought if we mean by language the eidetic system of "expression", expression as it is essentially different from sign.

We can introduce mental experiences to clarify this subject. Imagine that you want to solve a geometrical question. You review the problem, then you reach a conjecture, as if you recognize an implicit relation between two elements, then you say it to yourself in words; but you have had an experience of what you have expressed and you *knew* it before that you state that sentence in your mind. And you can even leave that sentence when you begin to say it. The sentence can have different grammatical structures and even can be stated in different languages if you were bilingual, or were fluent in several languages; or even you can forget a word and it take a while to find a suitable word. Trying to realize this variations, one can experience that the content remains same as it has been already there.

Then why we use words if we can have wordless expression? Why all these verbal speech in our mind? To give a possible answer to this question we should take into account psychological issues not merely logical ones, since while the bare expression is in a close relation to the meaning and thus belongs to the ideal sphere of logic, verbal statements, being the system of indication which operates due to "motivation" in mind[39], widely depends on social and psychological

---

[39]Husserl observed the fact that indication is based on mental motivation: A indicates B if the belief in A *motivates* the belief in B.(Husserl, 2001b, p. 272)

affairs.

But in so far as we can control our minds it can be said that we most of time repeat our thoughts in verbal form to ourselves perhaps because of one of the following reasons:

- We imagine ourselves asserting the current thought to the public or to certain persons and also imagine their reactions, possible critics or requests of explanation. By this mental method we revise our manner of stating or even we decide about that which theme is more important and in which order the argument should be present in order to be more resultful and better understood. This act may effect the rest of our thinking and be determinative, to some degree, in arising the other original expressions.

- To review a thought, i.e. an expression, by means of indicative language may have a significant role in *memorizing* and *remembering*. (later we will see that a mere perception is not reviewable except on the basis of expression of it.)

- It is possible that because of some psychological facts we can not stop the stream of speeches in our minds, speeches that have been heard once or more, have become important for us, are recalled frequently due to unconsciousness affairs, or are recombined spontaneously. Thus we decide to capture our mind by directing the mentioned stream by the speech of our own expressions. It can be considered as a kind of self-training which a philosopher or a scientist may do more than other people.

Perhaps other alternative can be mentioned to explain the presence of verbal speech imaginations in mind. However, whatsoever can be said about this issue, it is related to the nature of indication and its bases like motivation or association of believes and so on and has nothing to do with the nature of expression. What is of importance for us in this investigation is the possibility of inner dialogue but it is not crucial whether this dialogue can be verbal or not.

However we can see that Husserl could believe in the possibility of genuine inner dialogue if by dialogue we mean only expressions. On the other hand it is also possible to have inner verbal dialogue but it will not be genuine expressions, rather it may or may not be the effect of a genuine expression.

1.4.3 Three accounts of language

Considering the previous explanations it can be said that there are three senses in which we may speak of "language"; and these three levels may be occasionally, in the philosophical or scientific literature, spoken of without paying enough attention to their difference.

1. At the deepest level we have a faculty of the Ego by which he or she is able to express, and thus conceptualize the objective and make it understandable. This is the faculty of shaping meanings and facing them as such.

2. The realm of expressions, as products of the acts of expressing—not as the acts themselves—, has its own structure; and this structure is what is called Pure Grammar. We may have expression before indication and this is what makes indicating and using meaningful signs possible. Then the Pure Grammar has nothing to do with signs; it concerns the *a priori* properties and relations of meaning.

3. We have also the system of signs which function to indicate meanings. This system is what is called public language or communicative language. "Language" in this sense refers to the structure of signs which stand for meanings—not to the structure of meanings as such.

Consequently, one can figure out that a philosopher, while speaking about language, has one of these three levels in mind and takes it as the definition of language and then attributes his or her own idea to what he or she has taken to be "language". Considering this point, we can argue that whether or not the contributions of a philosopher to the subject, is compatible to the phenomenological framework and can be considered as a potential development of it in a specific direction or not.

Just three important examples:

For Brouwer (Brouwer, 1996) the character of language is its role in communication and memory. Then it can be seen that he speaks about what we distinguished as level 3. His viewpoint is that thought and knowledge in general and mathematics in particular are originally languageless.[40] Now his remark concerns the public language; that is completely acceptable in the mentioned phenomenological system, for thinking is independent of a given language.

However, when Levinas says that:

> Language does not exteriorize a representation preexisting in me: it puts in common a world hitherto mine. Language effectuate the entry of things into a new ether in which they receive a name and become concepts. It is a first action above labor, an action without action...
> (Levinas, 1979, p. 174)

he speaks about the first level; and thus his remark is also acceptable from the phenomenological point of view. Then there is no conflict between Brouwer's belief in languagelessness of thought and that of Levinas in considering concepts as language-based, rather they speak of two different things.

On the other hand when Wittgenstein speaks about language as if it is a system of signs obviously he has an account similar to what

---

[40]Brouwer (Brouwer, 1975) frequently emphasizes this remark. For a comprehensive study see (van Dalen, 1999).

1.4. DESCRIPTIVE PHENOMENOLOGY OF EXPRESSION

we have distinguished as level 3. Nevertheless as he gives priority to this "language" over rule following, recognizing or even thinking, his standpoint is not acceptable according to phenomenology.

Of course because of the ambiguity of the word language it is possible that a researcher does not keep a same account in the process of his study. Then such an study will be hardly useful in comparison to those investigations in which one keeps his account in a certain level (which follows from the insight about the nature of the subject except that there are still lack of clear terminology). Such an investigation can be basically accepted or rejected and in the former case regarded as an acknowledged contribution to the subject. That is to say, before evaluating a theory about language we should determine about which "language" it is supposed to discuss.

Considering these three levels, we can define certain sciences in regard to each of them. First of all we should mention the fact that, by distinguishing between expression and indication, we are able to notice not only the peculiarities of expression but also the peculiarities of indication as such. If we keep the notion of meaning apart from the process of transmitting from the sign to the signified, we can grasp this relation in its own home, i.e. psychological realm or the realm of the believes, and do this without worrying about imposing psychological matters to the meaning. Husserl has found the root of indication in the relation of motivation in human mind. He says:

> we discover as a common circumstance the fact that certain objects or states of affairs *of whose reality someone has actual knowledge* indicate to him *the reality of certain other objects or states of affairs, in the sense that his belief in the reality of the one is experienced* (though not at all evidently) *as motivating a belief or surmise in the reality of the other.*[41] (Husserl, 2001b, p. 270)

---

[41]In ihnen finden wir nun als dieses Gemeinsame den Umstand, daß ir-

Human being could recognized this relation and employ it in regard to recall a belief or a notion in one's mind, including that of oneself. It is natural to suppose that at first human beings were employing those items that are naturally motivating, but after that they could train themselves to use artificial motivators. Of course we assume the faculty of knowing has priority, so that on the basis of it human can be, relatively, aware of its potentialities. We are not to claim that system of indications can work without dealing with meaning in the sense that a being who has not consciousness of meaning can invent a system of signs. All we are asserting is that sign, qua sign, need not to essentially stand for meaning, but it can. Therefore the mechanism of signifying has its own right to be studied and on the basis of such studies we can go further in investigating the function of signs as they work for expressing in the public language.

In the following we attempt to define sciences belonging to the different aspects of language, considering the above mentioned distinctions between those equivocally called language.

- There is a place for the phenomenological investigation pertaining to expression and its constitutions, namely to peculiarities of the process of expressing and meaning-formation.

- There is a place for a science investigating on the nomological , a priori rules of meaning, which is called by Husserl **Pure Grammar**; and he has taken primary steps in this respect and introduced some seminal ideas in the forth investigation of his LI.

- The structure of signs, as such, is a specific field that a specific

---

gendwelche Gegenstände oder Sachverhalte, von deren Bestand jemand aktuelle Kenntnis hat, ihm den Bestand gewisser anderer Gegenstände oder Sachverhalte in dem Sinne anzeigen, daß die Überzeugung von dem Sein der einen von ihm als Motiv (und zwar als ein nichteinsichtiges Motiv) erlebt wird für die Überzeugung oder Vermutung vom Sein der anderen. (Husserl, 1913, p. 25)

1.4. DESCRIPTIVE PHENOMENOLOGY OF EXPRESSION

science may be appropriated to it, hence **Semiotics**. "Semiotics" and "semiology" are often used as synonyms, but in regard to our intention to the issue it is better to notice the difference. We shall define them as following: Semiotics should deal with *eidetic* rules of signifying and ideal relations of signs. **Semiology** is the name for the experimental studies of the *actual* use of signs in societies and cultures.

— **Linguistics**, therefore, is a branch of semiotics which deals with a specific kind of signs, namely locutions, with a specific function, namely to signify the meanings.

— A particular science can be introduced concerning precision in transmitting the items from the thought, from the realm of meanings, to the communicative language. This science should deal with the manner of *precisely* indication of the meanings. We shall call this science, in accordance with Husserl, **Apophantics**. It is, of course, a task of semiotics that keeps signs apart from ambiguity, but there are some special problems while dealing with meaning as that to be signified. Apophantics should benefit from the semiotics and linguistics, from one side, and the theory of pure grammar, from the other side. The importance of apophantics will be more clear, more we know about the peculiarities of pure grammar and special difficulties belonging to the sphere of the meaning.

The achievements of apophantic studies should be used to realize the Leibnizian idea of "Universal Characteristic". That to reach a final universal characteristic seems to be impossible should not imply that any attempt in this respect is useless. Indeed we believe that the ultimate universal characteristic is practically impossible, because it needs a complete knowledge of pure grammar from one side and having a proper, perfect system of signs from other side. However, we can try to make our language precise in every phase, in regard to our achievements in pure grammar and semiotics, and this precision in its turn will help us in making the investigations and argumentations about the subject-matters more perfect.

1.5 Constitutive Phenomenology of Expression

By the distinction we have explained above, Husserl mainly intends to grasp the notion of expression in its purity. Having kept apart the notion of expression from certain usual confusions, a phenomenologist becomes more able to face the central question: what is the peculiarity of expression?

We have learned that it is not indication and we should keep it apart from the stuff related to indication (like motivation, association etc.) and this observation helped us to know what is not expression. But yet what is the own function of expression?

In *Ideas I* Husserl writes:

> 'Expression' is a remarkable form, which allows for adopting to every 'sense' (to the noematic 'nucleus'), and raises it to the realm of 'Logos', of the *conceptual*, and thus of the *'universal'*.[42]

Husserl says that perception is not an expressive act; but thinking about it, remembering it, meaning it and so on are expressive acts. Through the act of expression the matter of the perception is conceptualized and thus brought to the realm of logic and knowledge. There are a lot of perceptions in every day life each of which can be considered from various aspects. But not all of them are actually thought about or recognized as a unity of a mental state. The latter

---

[42]"Ausdruck" ist eine merkwürdige Form, die sich allem "Sinne" (dem noematischen "Kern") anpassen läßt und ihn in das Reich des "Logos", des Begrifflichen und damit des "Allgemeinen" erhebt. (Husserl, 1976, p. 286)(translation is mine

1.5. CONSTITUTIVE PHENOMENOLOGY OF EXPRESSION

is happen by the expressive act of the ego. By expression they are distinguished from the *hyle* of affection, and raised to the realm of thought, and grasped as conceptual. The peculiarity of expression should be studied in regard to this process.

It should be clear that without meanings we have no "Object", in proper sense of the word, but only the stream of experiences. These experiences, once conceptualized, can be referred to in terms of objects and states of affairs, but before that we can not indicate them or take them as identical; we can only experience them. However, it is the characteristic of the ego, of its consciousness, that it can turn toward the experiences through its acts, and thus objectify them, or rather raise them to the realm of Logos. Now the function of expression should be found in this stream of acts.

What is the relation between meaning and expression? How does expression raise an experience-content to the realm of conceptual? What does it mean to speak about the realm of Logos, of the conceptual, and what is its place in the life of the ego? These are the questions we depart to study in this phase of the investigation.

As a starting point, it maybe thinkable to assert that, as it seems to be implied from Husserl's above mentioned opinion, the expression is the dynamic version of meaning, or expression is meaning as it goes beyond itself. It means that expression is not a separate entity which must be in a specific relation to meaning or refer to it, rather it is a process by which meaning manifests itself. However, we think that developing this idea will not be satisfactory; for a lot of questions will remain untouched: the origin of the meanings, the procedure of conceptualization, the how of relation between expression and indication. But there is an insight in this idea: to consider expression as process. We suggest trying this idea but in the reverse way. It means that may be it is not the case that meaning goes beyond itself and causes expression but expression is an act which is called meaning in regard to its product, irrespective to its dynamism. Let's examine this idea.

The main question that we face in the beginning of our way in developing the mentioned idea is this: If expression is prior to meaning and if to express is not basically to express the meaning, then what is that to be expressed? The answer should be asked for in the realm of consciousness and experience while not yet equipped with meanings. In the pure realm of consciousness we have intentions and also (not-objectivated) perceptions. Now we should see that what is the difference between mere having an intention or mere having a perception and expressing it.

Assume that we intend toward something in a certain respect, say toward a fruit in respect of hunger; now that phenomenon may or may not fulfill our intention. The point is that we can have another intention toward our previous intention, and this is the peculiarity of consciousness; on the basis of this peculiarity and on the basis of this very latter intention we can be *aware* of that whether or not our former intention has been fulfilled, or in better words, whether or not that phenomenon has fulfilled our intention. Therefore, we will have a secondary intention—which is supervenient on the primary one—toward the same phenomenon. Assume that the mentioned fruit would be really eatable. Hence, having consciousness of our hunger-intention being fulfilled, we can have another intention toward that fruit as fulfilling our hunger-intention. Thus, that phenomenon turns out to be an object of knowledge; and, for the first time, the intentional unity directed toward that phenomenon constitutes its essence as an identical object. Just like any intention, this secondary intention may be not fulfilled. It means that its possible that we consider something as eatable while it is not so. However this fact is recognizable on the base of another higher intention. To explain, we can intend toward something as fulfilling our hunger-intention, regardless that we are hungry or not i.e. regardless that we have hunger-intention at the time or not. It is possible that that thing not be eatable but we consider it as falling under the concept of eatable, or our conception

1.5. CONSTITUTIVE PHENOMENOLOGY OF EXPRESSION

of eatable is so that covers that particular thing. This would be a mistake, a correctable mistake, and correctable on the basis of another intention which, on the basis of the intention constituting the concept at work, intends toward this or that phenomenon as instance of that very concept.

Let's articulate this process:

1. We intend toward objects in a certain respect. We intend toward objects, which is not yet considered as Objectified and Objectively separated, through this or that intention.

2. We can also intend toward our intention in phase 1. This intention calls our attention to the relation of that lower intention with the objective, and thus to the issue that whether or not that intention is fulfilled.

3. Thus we can also intent toward the objective again, and this time not through our first intention but above it and through the new intention which intends toward the phenomenon as correspondent to the previous, now objectified, intention.

Now let's come back to the expression. We have act of expression in the third phase of the above described process, wherein our intention turns toward outside the interior realm, for in so far as we have simple intention, say hunger, and simple perception or experience, we are still in "inside", in the realm of "immanence". A phenomenon which is the subject of such intention may simply fulfill it or not and there is not yet any consciously attention to the phenomenological "outside". However, when we become aware of our intention and turn toward it as the subject of another intention, we face the phenomenon from distance and as "external", i.e. external to our former intention, which may or may not fulfill it. Now we can intend this phenomenon to fulfill our former intention. This latter intention is not the same as the former one, but supervenes on it: To intend to eat something is not the same as to intend something to be eatable. Expression appears here: the attention of the ego from the immanence (mere intentions and experience) toward outside by means of passing from the former intention to the latter one, by means of intending some phenomenon which is essentially "distanced". This act of exiting from the realm of a mere intention and then intending the phenomenon as transcendent to the primary intentional experience, but still correlated to i,t is called expressing.

Now, if we consider what is expressed, as expressed, in its inert identity not in action, we will have *meaning*. This meaning is objective and has its objective place, still in consciousness, for, by expressing, the ego does not exit the realm of consciousness: though it intends the external as external, but the act expressing itself remains immanent. Thus the product of expression, namely meaning, is *in* consciousness, and this is not in contradictory with its being objective because this being in inside is not natural one and completely based on transcendentality of the ego, namely its ability to iterate through intentions. To better explain this situation, in the following, we will benefit from noetic-noematic[43] analysis and attempt to illustrate the unclear aspects of our proposed idea.

Studying the issues related to the realm of pure ego and its experiences, Husserl distinguishes two different aspects of experience. One of which is the stuff of experience or sensory data, while the other is the quality of intentionality through which the ego holds in touch with that stratum of primary content. Husserl names the first phase *sensile hyle* and the second *intentional morphe*(Husserl, 1969, p. 247). In order to be more specific, while speaking of this distinction, Husserl uses the term *noetic* to indicate the phase of intentionality in every experience. Therefore:

---

[43]Roughly speaking, the noema is the sense-content of every experience and is in correlation with phases belonging to every act of experiencing as such. Noema is different both from object and from meaning. This is not an uncontroversial stance regarding to different interpretations of Husserlian doctrine of the noema. I will return to this issue in the next chapter.

1.5. CONSTITUTIVE PHENOMENOLOGY OF EXPRESSION

The stream of phenomenological being has a twofold bed: a material and a noetic.[44](Husserl, 1969, p. 251)

Since the difference between those we have described in the process of expression as first and second intention, namely to intend toward a phenomenon in a certain respect and to intend toward a phenomenon as fulfilling the former intention, is very important to understand what we want to say about the origin of expression, we intend to reexplain that difference in terms of the above mentioned distinction in the realm of experience.

One may ask what is the difference between to intend to, say, touch something and to intend it to be touchable. Why we should not consider these two cases as two modes of statement rather than two different acts? We can explain our stance by an example:

Assume that we have an intention toward beauty. Now in the stream of experiences there are sensory data of colours and smells and so on. There are flowers that we see or songs we hear, but notice that we have no concept and no word yet and we do not discern between objects but we can sense the data through the currency of experience. Now through our beauty-intention we intend toward a piece of this stream, say toward a flower, then our intention may be fulfilled or not. In any experience of beauty we have an intention, a noetic phase, and a material phase, say colour of a flower, which more or less fulfill this intention. Without this intention there would be no beauty. This intention allows us to enjoy beauty but it is not sufficient to expect some phenomenon to be beautiful. To intend toward something as beautiful is a noetic act which is built on the noetic act of enjoying beauty. And the material of experience of distinguishing some phenomenon as beautiful is not a spontaneous sensory data but the phenomena as regarded from the above and in correspondent with the act of enjoying beauty.

If the ego had not the ability of going higher, it only would

---

[44]"Der Storm des phanamenologischen Seins hat eine stoffliche und eine noetische Schicht."(Husserl, 1976, p. 196)

experience the data and during this stream it would sometimes enjoy and sometimes not. But the scenario does not end here. Although receptivity is a vary important feature of the ego, it is not restricted to it. Thanks to the transcendental consciousness it can turn toward any phenomenon even its own acts.

Considering the fact that expression is a higher level noetic act, it should be clear that it is not a mere duplicating. Rather, by means of expression, the ego reaches a peculiar relation to the phenomenon which is not present in the non-expressive relation. When we constitute a flower as beautiful or a fruit as eatable, we rise from an immediate and dull relation to a mediated and dynamic one. Therefore, we should never consider expression as a transcription or duplication. We do not deal here with doubled intention or doubled phenomenon. Rather, we have in this act a supervenient intention and an expressive entity, namely *meaning*.

Husserl says in his *Ideas I*:

> The layer of meaning is not, and in principle is not, a kind of reduplication of the lower layer.[45]

Also in a text related to the revision of the sixth investigation of the *LU* Husserl says:

> But our study shows clearly that for all intuitive synthetic notions and about all their formations etc. there is a mental expression, but it is more than a repetition, an imprint and the like., but it is a comprehending expression, a "cognizant"...[46]

---

[45]Die Schicht des Bedeutens ist nicht, und prinzipiell nicht, eine Art Reduplikation der Unterschicht. (Husserl, 1976, p. 291)

[46]Aber unsere Betrachtung zeigt deutlich, dass für alle intuitiven synthetischen Meinungen, über alle ihre Formungen etc. hinaus liegt ein geistiger Ausdruck, der aber mehr ist als eine Wiederholung, ein Abdruck u. dgl., sondern esist ein begreifender Ausdruck, ein "erkennender"... . (Husserl, 2005, p. 231)

1.5. CONSTITUTIVE PHENOMENOLOGY OF EXPRESSION

Now, considering the meaning as the correspondent product of a primordial act of expression, we will be able to better understand the peculiarities of meaning and its relations. An important point is that, according to our analysis, there is no need that there *exists* a phenomenon which we speak of its meaning. Regarding to the distinction between primary un-expressive intention and the supervenient expressive intention, we know that in order to intend something as the fulfiller of our primary intention it is not required that our primary intention be actually fulfilled. One can become hungry, have hunger-intention and be aware of his or her intention, then one can have another intention toward phenomena as fulfilling the hunger-intention, while has not yet eaten anything and has no idea of the quiddity of what may obviate his or her hunger. It means that one can have a concept of food without having an experience of eating. Now it is possible that due to further attempts one obtains food and that primary intention fulfills. For some intentions, it is possible that one concludes that there is no phenomenon fulfilling that intention. Suppose that one has an intention to be transparent. Let's assume that there has been formed such a meaning and it has been called by a name, say, *transparenter*. Here we deal with a meaning, but our experiments shows that there is no instance for such a meaning, for such a concept[47], there is no phenomenon fulfilling the intention of being transparent. We may even show a priori that it is impossible that there exist a phenomenon fulfilling a given meaning-intention. But none of these makes it "meaningless".

---

[47]Albeit, concept and meaning are not the same, concept is one category of meaning; the others are proposition, interjection etc., corresponding to the different forms of experience-expressing. We will elaborate this point in the next chapter.

1.6 Transcendental Possibility of Inner Dialogue: Inner Alterity of the Ego

Given that it is possible, according to phenomenology, to use language in solitary life —as we were trying to demonstrate it until now— we face another important question. To have inner dialogue in proper sense in addition of having inner language needs that there be a kind of duality. Is there any duality for subject in itself? We try to defend the affirmative answer in this section. By doing this, we hope that, the roots of possibility of inner dialogue will be clear; and we will be permissibly assert that inner dialogue according to phenomenology is not only conceptually but also transcendentally possible. It means that it has its foundations in the transcendental intentionality and thus constituted in a basic level of ego's activity.

In the very beginning of accomplishment of the phenomenological method, namely in the process of epoche, I confront a kind of split in myself:

> If the Ego, as naturally immersed in the world, experiencingly and otherwise, is called *"interested" in the world*, then the phenomenologically altered and, as so altered, continually maintained attitude consists in a *splitting of the Ego*: in that the phenomenological Ego establishes himself as *"disinterested onlooker"*, above the naively interested Ego.(Husserl, 1960, p. 35)

Therefore there emerges a distinction which makes the ego able to bracket his own natural motivations, interests and relations, and thus objectify itself and in a higher level constitute itself.

That is to say, it is not the case that "I" am simply an entity with certain relations with others, rather I find myself as possessing genuine existence, and not only find myself so but also constitute

1.6. INNER ALTERITY OF THE EGO

myself and as a dynamic identity continuously become aware of myself.

> The ego is himself *existent for himself* in continuous evidence ; thus, in himself, he is *continuously constituting himself as existing.* (Husserl, 1960, p. 66)

The ego is aware of his/her being and his/her acts and states. However, the important point is that the ego is *active* in self-conceiving and that it is the *possibility* of being conceived that essentially belongs to every cogito not the *actuality* of it.

> It pertains in general to the essence of every cogito that a new cogito of the kind called by us "Ego-reflection" is in principle possible, one that grasps, on the basis of the earlier cogito (which itself is thereby phenomenologically altered), the pure subject of that earlier cogito.[48] (Husserl, 1988, p. 107)

Thus pure ego not only faces itself as an entity in the world and goes higher but also faces itself as pure ego.

> It consequently pertains, as we can also say (since the same obviously applies to this reflective cogito as well) to the essence of the pure Ego that it be able to grasp itself as what it is and in the way it functions, and thus make itself in to an object.[49] (Husserl, 1988, p. 107)

---

[48]Zum Wesen jedes cogito gehört es generell, daß ein neues cogito von der Art, die wir "Ich-Reflexion" nennen, prinzipiell möglich ist, welches auf Grund des früheren, sich dabei phänomenologisch wandelnden, das reine Subjekt desselben erfaßt. (Husserl, 1952, p. 110)

[49]Somit gehört es, wie wir auch sagen können (da auch für dieses reflektive cogito natürlich dasselbe gilt), zum Wesen des reinen Ich, sich selbst als das, was es ist und wie es fungiert, erfassen und sich so zum Gegenstand machen zu können. (Husserl, 1952, p. 101)

The Ego can confront itself and posit itself as object. In accordance to these remarks, we should note that it belongs to the nature of consciousness that it is not the case that everything in that realm are highly clear and without shadows. But the boundaries of consciousness are less clear and the ego with its concentration becomes more aware of them. Husserl differentiates between *alert* and *dull* consciousness and describes the evolutions in consciousness in a dynamic manner (Ideas II §26). It can be said then that every consciousness is able to be subject of a self-clarification and, in this sense, it can be also a moment of self-determination.

Therefore, we see that self-objectifying is a very significant theme in the transcendental phenomenology. Consequently one can speak about self-determination, self-training (Ideas II §58), self-criticizing and also self-questioning. And perhaps the very important notion of self-responsibility should be interpreted as a less metaphorical term and we know that how this notion is principal to begin the philosophical effort in Husserl's eyes.

The realm of the ego is not static and homogeneous. Ego confronts himself and objectifies himself from certain aspects in each act. Since the act of objectifying is interwoven with conceptualization and since, as we saw before, to conceptualize something is a kind of expressing it, the ego deals with itself through expressions.[50] Every act can be conceived by another act (Ideas II §23) or in other words every act of the ego is in principal expressible (LI Investigation Six §2). And thus ego is continuously confronted with itself and criticizes itself and goes higher from previous cogitos through expressions.

This is a peculiarity of the transcendental ego: the possibility of reflection and accordingly that of *iteration* in which the ego becomes both subject and object while maintaining its identity. Sebastian Luft (Luft, 2002) discusses the importance of these peculiarities for

---

[50] Albeit, we have also non-conceptual expression and it is also very important in respect with self-encounter.

1.6. INNER ALTERITY OF THE EGO

the phenomenological method; and there he rightly remarks:

> How is this objectification of one's subject while maintaining its unity possible? Husserl tries to find the solution with his doctrine of the ego-split.[51]

Ego-split, thus, becomes a very important theme in respect with the phenomenological method and the process of transcendental reduction. However, as Luft explains, this ego-split which is understandable in the light of transcendental attitude should not be confused with some possible natural split in one's identity; nothing psychological is here at work while speaking of reflection and self-objectifying.

Reflection is an experience of a higher level. In reflection on an act or on an experience of a phenomenon, the ego transcends that phenomenon and correlatively transcends itself as having that experience; so in the higher level it maintains as ego-subject while turning its regard to the ego-object. This ego-split becomes a ground for another meaning of ego-split which is more important in respect with the inner alterity. The important notion here is the *horizontality* of intentions. In each experience there are other possible intentions which makes further experiences of the different aspects of the same phenomenon possible. Such an horizontality itself becomes manifest and turns to be a theme in the higher experience of reflection. In reflection on a phenomenon the ego not only attends to its actual, i.e. *positional*, experience, but also to the possible experiences which could be made from other perspectives. In a tight relation to the reflection are the notions of *quasi-positional* acts, *fantasy* and *as-if* (*als ob*) experiences, and all of these are related to the notion of ego-split in which the ego fantasizes itself as perceiving the phenomenon from the other perspective so that the object-ego is not necessarily

---

[51]Wie ist diese Verobjektivierung des eigenen Subjekts bei gleichzeitiger Beibehaltung seiner Einheitlichkeit möglich? Die Lösung versucht Husserl mit seiner Lehre von der Ichspaltung zu geben. (Luft, 2002, p. 111)

that having the positional acts but a *modificationl* ego having some quasi-positional acts. Husserl discussed these topics in details in the paragraph 44 of the second volume of the *First Philosophy* (Husserl, 1959, pp. 112–120).

Therefore, on the bases of the ego-split and also that of the horizontality of intentions, we have also the ego-modifications. The modified ego could not have a *primordial* experience; nevertheless the modification of the ego may be itself a primordial experience. I discussed earlier that the constitution of the meanings depends on the primordial expression, however once a meaning is constituted it can be used in a non-primordial way. Therefore the quasi positional acts of the modificational ego may be expressive and thus may use of the already formed meanings. That is to say, the inner interaction which takes place in some acts of the reflection, may be expressive hence dialogical. Although the modificational ego does not bring about genuine evidence so does not perform genuine judgments[52], however as it can use the meanings in a non-primordial way, it can express a proposition to make an as-if judgment or to assert a modificational judgment and this is the way in which the inner dialogue which takes place during some reflections goes.[53]

The inner dialogue itself is a primordial experience since the outer dialogue itself in certain respects constitutionally depends on the inner dialogue. This is so because the constitution of the external other is partially based on the inner alterity.

Husserl frequently asserts that the Other is constituted in the ego:

> [Other] is therefore conceivable only as an analogue of something included in my peculiar ownness. Because of its sense-constitution it occurs necessarily as an "inten-

---

[52]About the relation between evidence, judgment and proposition, I will discuss in more details in the next chapter.

[53]The role of [inner] dialogue in the logical reasoning and the contribution of dialogical explanations to logical problems will be discussed in chapters 3 and 4.

tional modification" of that Ego of mine which is the first to be Objectivated, or as an intentional modification of my primordial "world": the Other as phenomenologically a "modification" of myself... (Husserl, 1960, p. 115)

Nevertheless, this issue remains somehow controversial among scholars. Zahavi's short but brilliant paper (Zahavi, 1997) about this issue is quite helpful. He shows there that how the horizontality of intentionality, which is one of the significant observations of phenomenology, introduces the concept of *open intersubjectivity*. That is to say, the notion of intersubjectivity which represents the existence of various standpoints about every objective identity is properly constituted in the ego. On the basis of this open intersubjectivity the constitution of others is possible, as the beings which have their own standpoints and are essentially perceived as possessing their proper perspectives.

Therefore the otherness in the ego itself is the foundation for the concrete intersubjectivity. Although, and this is a very important point, the latter should be by no means reduced to the former. Zahavi admirably discusses this point (Zahavi, 2001, p. 203).

Consequently considering Ego's inner non-homogeneity and its inner dynamic alterity and in regard to its being equipped with subjective language in order to objectify, conceptualize and clarify objects from different aspects and even from different standpoints, we can see the place of inner dialogue in a deep level of efforts pertaining to knowledge i.e. thinking and analyzing in order to illuminate the phenomena and to grasp things as such.

1.7 Reply to Wittgenstein's Private Language Argument

Now, we are to reply to Wittgenstein's objection to the possibility of inner dialogue.

Considering the different accounts of language which we discussed in the section 1.4.3 and the ambiguities around the concept of meaning which has been shown in section 1.4.1, now we can see that Wittgenstein's argument is based on a narrow account of language. Thus it involves certain confusions about the meaning. Particularly, Wittgenstein takes language as an essentially symbolic system, while this is only one of the three accounts of language in the context of both ordinary and philosophically speaking of it. Wittgenstein has this presumption since his early period and it remains same where he revised his main thoughts.[54] The difference between the two periods is that in his early stance language is a system of symbols which are supposed to picture the state of affairs and this system could work also in solitude, whilst in his latter stance language is a system of signs which refer to each other and their functioning is essentially realized in the interactions in community. He argues for this latter stance and brings arguments to support it, including argument against private language.

After this general remark, we mention the premises and phases of the private language argument and examine them by means of the hitherto explained phenomenological remarks.

Considering our reading in section 2 we can say that the elements of the argument is as following:

α) A sole person could not follow a rule. (Rule-Following Argument)

β) The word is in fact replaced in lieu of (natural) displaying of sensations. It does not originally introduced as signifier to meaning or expression. (Substitution Argument)

---

[54]Precisely speaking, the early Wittgenstein approves that "sign" is arbitrary. Nevertheless, he distinguishes between "symbol"—which, as he puts it, is equal to expression and is the essence of language—and "sign" which is the physical sign standing for the symbol. See propositions 3.31, 3.321, 3,322 and 3.323 of Tractatus (Wittgenstein, 2015).

1.7. REPLY TO WITTGENSTEIN

γ) It is more likely meaningless to speak of "private"; we have no fundamentally private experience about which we are in need to have language. (Private Object Argument)

δ) It is impossible to establish a language exclusively by means of assigning signs (words) to things and states. (Stage-Setting Argument)

The items α, β and δ together result in impossibility of having a language in solitude; and thus the originality of inner language rejected. The items γ and δ together result in impossibility of having a language in order to speak of private experiences. (see p. 19)

Among the aforementioned premises and phases, we approve that the item δ is true, but it is inconclusive unless put beside others. In the following I want to criticize items α (1.7.1), β (1.7.2) and γ (1.7.3), and show that the item δ (1.7.4) is ineffective. Thus I will reject the Private Language Argument in its both cases, i.e. as rejecting the genuineness of soliloquy and as rejecting the language for the private.

1.7.1 Rule-Following argument and the place of the Meaning

A crucial step of the private language argument is to show that to create a language by mere denoting is impossible. In other words Wittgenstein tries to show the flaw of descriptive theory of meaning by showing that to fix indicators is not as simple as it may seem, it needs and presupposes some abilities: the ability of rule following, recognizing something as identical, etc.

However, the more Wittgenstein truly shows the complexity of the problem, the less he considers the intricacy of consciousness and abilities of mind. He does not regard the distinction between intention and fulfillment and accordingly he does not notice the distinction between meaning and sense. He sets a scenario in which concepts must appear in the unity of perception and every meaning must refer to something external and stands as indicator for it. If someone defends such a standpoint, like Wittgenstein himself did in his *Tractatus*, then private language argument, at its best, shows that *that* standpoint is not acceptable, but this argument does not go far if considered precisely.

Wittgenstein's reliance on the Rule-Following argument, that is a person considered in isolation can not follow a rule, is based on some confusions. Firstly he does not take to account the intentionality of consciousness, thus he could not distinguish between intention and fulfillment. This confusions reproduces itself in the Private Language Argument.

If we distinguish between intention and fulfillment, Wittgenstein's argument can no more violate the ability to recognize, to identify and to use signs by a lonely subject. As we saw he says:

> if we construe the grammar of the expression of sensation on the model of 'object and name', the object drops out of consideration as irrelevant.

His alternative is to consider language as constructed on the model of words and social interactions. However, he should have considered the model of intention of an object, the act of expressing and objectifying based on this intention instead of the model of object and name.

As Wittgenstein puts it, in private life, meaning must be originated in the perception or in the naive being confronted with the object, but perception alone is not sufficient to establish language. Then, in Wittgensteinian point of view, in solitary life we have neither private language nor meaning. However, being in agreement with the second premise above, one can ask why we should take for granted that in solitary life meaning has no choice but to arise from perception.[55]

---

[55] Of course Wittgenstein himself once was defending such a point of view,

1.7. REPLY TO WITTGENSTEIN

Phenomenology makes this observation more clear that to mean something is not identical with to perceive it and thus meaning unlike perception needn't to represent or refer to something in the real world. Therefore meaning in its nature differs from perception and hence the peculiarity of meaning should not be reduced to that of perception which is the representation, or so to speak, mirroring the objective external world. Accordingly while the genuineness of perception is interwoven with the concept of truth and restricted to it, meaning of an utterance does not depend on the truth of it. In particular, to intend some act is different, and has its different origins, from to do it rightly.

Actually, Wittgenstein argues about the private language as if expressions must operate for senses (e.g. a specific sensation S). But, expressions are in close relation to meanings. Thus, firstly, an expression (and its indicator say sign 'S') should stand for the (ideal) *meaning* corresponding the mentioned sensation, not for its circumstantial occurrence or the sensile stratum of the relevant experience. Each time sensation S occurs one can consider it as identical with previous ones, for the meaning-intention remains same. There intention exists as criterion for the relevancy. On the bases of the identical meaning-intention one can speak of fulfilling or not fulfilling during the different experiences; so that one can say that this is S or not. (The fact that one might be mistaken is not

---

following a Feregeian metaphysics and in close accordance with descriptive meaning theory, and by these mentioned arguments he criticizes his previous beliefs.

Related to this is the Leibniz-Locke debate about the content of thought, about which I side myself with Leibniz. However, regardless to this debate, even in the external perception we are not quite passive, rather every content of perception is what it is thanks to an intention which makes it distinct. Therefore the moment of intending and that of fulfillment are different, even if they were co-existent—which is not always the case. In the distinction between intention and fulfillment, meaning is related to the former. So any analysis which ignores the former and sets forward an argument on the assumption of rendering the meaning to the fulfilled perception is just inappropriate.

relevant to our current investigation.) Therefore one *can* follow a rule regarding the ideal criteria which are there as the meanings.[56] Therefore, it is not the case that for a sole mind "thinking one was following a rule would be the same thing as following it"[57], on the basis of which Wittgenstein concludes there is no true rule following in solitude.

Wittgenstein's account of rule following is not free from big deficiencies. If rule following would be impossible in solitude, it would be also impossible in community. Because rule following is based on, and the consequence of, free will. It would be very strange to say that a sole person has not free will but a community has, so that a person as a member of a community can be said to follow a rule.

If someone says that rule following has nothing to do with free will, it would be right to say that he does not speak about rule as such, rather he speaks about another issue. But if someone says that free will must be attributed to the community rather than the person, we admit that his position properly makes him able to reject the private language. Nevertheless, he would support a counter-intuitive claim by appealing to a much more counter-intuitive thesis. If someone says that rules must be determined externally for a subject of free will, we face a very prevalent mistake. We are not here to explain this moment, but, just to address it briefly, that one can adapt some rules or determine some external criteria as mediate points of checking does not mean that the rule as such must be external. The supposition that rules are external, that is, they are imposed as limitations to the free will from outside, is a category mistake. The rule as such is determined autonomously, or so to speak internally, by the subject.[58]

---

[56] As I said it is still possible that one committees an error in following a rule and does not recognize its error, but this does not mean that there is no criterion or no rule.

[57] (Wittgenstein, 2009, §202).

[58] Among the great thinkers who have elaborated this determinant point we

1.7.2 Substitution argument and the function of language

As we saw before, Wittgenstein in the fragment 244 of his PI speaks about expression such that it originates as social-based replacement for what he considers as natural expression. Geach (Geach, 1971, p. 121) properly describes Wittgenstein's viewpoint:

> "How that hurts!" or "it hurts here" would be Äußerungen of pain, serving not to express judgments in which the pain is described but as learned replacements of primitive (animal) pain-reactions. Similarly, "I am extremely angry with you!" is not a description of the degree of my anger but a replacement of primitive angry behaviour.

Wittgenstein explicitly says:

> The verbal expression does no more than some kind of wordless behaviour. (Wittgenstein, 2009, p. 231)

However, what does replacement mean here? When something "replaces" some other thing, everything must be remain same, for this is only a replacement. In the other words, we consider A as a replacement for B when in a given context A has almost every function that B had, though they are different regarding to their other properties. Therefore, the whole context is supposed not to be altered, while speaking of replacement in some aspects. It means that the modern human has no difference with the wild animal, because the anger of that animal is substituted with to utter "I am angry". But, if something changes by replacing A with B, then the function of A is different from that of B. What is the difference between saying that "I am angry" and executing angry behavior? If there is a substitution it does result in that these two are completely alike and there is no need for further studies.

---

can name Spinoza, Leibniz, Kant, Kierkegaard, Jaspers, and Brouwer.

We believe that in most cases there is no replacement, however even if there were a replacement it is irrelevant to the main study: expression (and language) has been constituted in other room and is chosen as substitute in certain circumstances. What is important for us is the problem of constitution of expression not that after being constituted what can be or can not be done with it. *Language, like any other instrument in human life, can have secondary uses.*

Indeed in most cases people uses locutions to *express* something not just substitute a brute act. However, even if people may use locutions to replace other reactions, that is they utilize some kind of speech act without intending to express something, though this may be of interest in sociological, psychological and anthropological studies, it is quite irrelevant to studies about the *constitution* of meaning and about the language as such.

It is clearly unacceptable to assert that: "The verbal expression does no more than some kind of wordless behaviour." This is a fruitless thesis whilst we are seeking for peculiar characteristics of language and its essential constituents.

Wittgenstein also brings examples of some expressions which are highly analogous to wordless behaviour, in order to point out that there is no sharp distinction between the two. We are in agreement with Wittgenstein that expressions like "Wow!" or "beam!" could be complete expressions; and there is no need to consider them as elliptical. There is no need to think that the complete expression should be a proposition so that every expression is an elliptical form of it. However, this attitude should not be extended so far that any kind of reaction, which is done by sound, is taken to be expression so that expression and reaction are considered as equivalent or as two sides of a same spectrum.

Wittgenstein's viewpoint is not the only to state that the proposition is not the main form of expression. Phenomenology admits this too. The disagreement is about the conclusion of this thesis: That expression is constituted not for representation but for its

1.7. REPLY TO WITTGENSTEIN

use in community (as a Wittgenstenian would say), or, that expression is constituted not necessarily for representation but as a peculiar intention of the transcendental ego toward the being (as phenomenology would say).

If we admit this latter idea and study expression as a peculiar act, we find out that there is no requirement that expression be totally representative, and even for the representative expression there is no need to be always propositional. Therefore, according to transcendental attitude the distinction between complete and incomplete expressions is not serious. Rather, an investigation may be dedicated to the general kinds of expression and accordingly the morphology of meanings and their synthesis.

Wittgenstein claims that the judgment is not the paradigm of expression. This is true, but this is not necessarily leading to the stance that expression is nothing but substituting some natural reactions.

Phenomenology could acknowledge most of insights behind Wittgenstein's discussion, but finds its overall attitude and its consequences groundless and inadmissible.

1.7.3 What is the private?

We should distinguish between three issues when speaking of a private experience:

1. Act of experience (Noesis)

2. The content of that experience (Noema)

3. The meaning (e.g. a concept) corresponding to that content which is an expressive and ideal moment (Noematic nucleus)

Wittgenstein does not notice these distinctions. It is true that each act of experiencing is real and objective and in this sense can not be considered as private. Wittgenstein's argument would be true if we had nothing but *act* of experiencing without its peculiar content and without the ideal meaning of that content. As if in the experience of pain or experience of being confronted with beauty, there are only the acts of experience. However, phenomenological analyses show that in correlation with the act of experience there is a peculiar, intuitive content of that experience, such as a special pain or a particular perception of beauty, which is not real—at least not in the sense that the act itself is— and its presence and effect in mind goes beyond the limits related to the event of the occurrence of that act. More importantly, this content can be a basis for other mental acts and above all for the expressive act which devotes a meaning to the content and thus bestows to it its identity.

Therefore, it is completely true that when I have toothache or when my hand hurts, this event—this act of experiencing—is a worldly act to which others can in principle have access, i.e. observe it directly or indirectly, get it as a subject of an experiment and speak about it. However, in each case, I have a peculiar pain, or perhaps suffer, to which no other person has access, no one is aware of its quiddity. If there would be a way to anticipate the quiddity of my pain for others, it is, first of all, by means of having access to the meaning of pain, not to the concrete event of experiencing pain by me.

If I have given meaning to pain, namely expressed it in a way, then each time I experience pain, despite their particular contents, I can recognize them as identical. But if there has not been such an expressive act, each experience of pain would give a distinct content; more precisely they can not be recognized as distinct or identical. Thus for me, and for any one who has no expression of this matter, there will be different experiences, though vaguely related to each other.

The claim of Wittgenstein seems to be that for a person considered in isolation every experience would be a different event and if there would not be an external criterion of identity, they can

1.7. REPLY TO WITTGENSTEIN

not be marked as identical. Wittgentein's argument could be valid if consciousness were nothing but the act of experiencing, namely mere noesis without noetic-noematic correlation, it means without intuition of the content of the experience and without ability of iteration in mental acts.[59]

1.7.4 Stage-setting argument and the distinction between expression and indication

The problem of saying, and that what is sayable, is one of the central themes of Wittgenstein's philosophical effort since his early work. In a letter to Russell he says:

> The main point is the theory of what can be said by propositions—i.e. by language—(and, which comes to the same thing, what can be thought) and what cannot be said by propositions, but only shown; which, I believe, is the cardinal problem of philosophy. (Wittgenstein, 1974)

Of course he has presented two radically different approaches to face this problem in two different periods of his life. However, certain accounts of language have survived in his viewpoint. While distinguishing between saying (*Sagen*) and showing (*Zeigen*), he never distinguished within the process of *saying* between *expressing* and *indicating*. In fact, he has used the terms *Sagen* (saying), *Ausdruck* (expression) and *Aussprechen* (utterance) somehow interchangeably. One can observe that for him saying was nothing but a special kind

---

[59]When I had already finished this writing, I learned about Meixner's recently published book, *Defending Husserl: A Plea in the Case of Wittgenstein and Company Versus Phenomenology* (Meixner, 2014), in which he sets forth certain phenomenological criticisms, mostly in the field of theory of consciousness and philosophy of mind, against Wittgensteinian viewpoints. In particular he rejects Wittgensteinian attitude toward the private object, and in this respect his arguments are related to our discussion.

of indicating, thus the difference between expression and indication remained undetectable for him.

If we take the peculiarity of expression into account, the problems about indication, denoting or ostension would loose their alleged importance. For the original issue of language is expression not indication. Thus that in some occasions to define something by means of indicating, that is assigning a sign to the object in question, is problematic has nothing to do with the essence of the language. From the phenomenological point of view it can be a hint to show that language is not rooted in indication. Only if one confuses between indication and expression, one would think that by rejecting the possibility of performing indication in solitude on the basis of Stage-Setting Argument the possibility of soliloquy in general will be rejected. The right consequence of such an argument is that indication can not be performed by an assumedly languageless sole person. Wittgenstein goes to say, thus language is social in its essence. A phenomenologist would see in this argument a support for his idea that indication is not capable to beget language.

The Stage-Setting Argument shows that language could not start by assigning signs to objects. Phenomenology and the latter Wittgenstein both approve that it is not the case that objects in the world are all in themselves distinct so that the function of language is to assign a name for each of such objects. Rather, they both admit, language itself has a role in making distinction among the objective; that is to say that the linguistic categoriality serves in the acts of objectifying and it is not possible to speak about the structure of the objects independent of the objectifying acts which are in close relation to language. The point of departure is about the account of the so-functioning language. There are different notions of "language" in ordinary speech; and the duty of philosophy is to keep a notion apart from ambiguity before ascribing some property to it. About language I tried to do this. Now it can be said that "language " in only one of its senses, namely in the sense of ability to *express*, is functioning in objectifying and making distinctions. Whilst , Wittgenstein's account of language is that of ordinary language, the third account in our analysis (sec. 1.4.3). Our disagreement is in that language in this sense can not have a role in the genuine thought, while in Wittgensteinian attitude this language is taken to be prior to the thinking subject and effective in thinking and objectifying. We admit that language in this sense can not be originated in solitude without a background of other language. However, we assert that language in its fundamental sense works also in solitude.

I hope that I have shown that having resolved the ambiguities around the notions at work it would be clear that the Private Language Argument is not conclusive; and although it is perhaps of some philosophical advantages as a reaction to Wittgenstein's own ideas in his early work, it does not introduce a generally reliable consequence.

1.8 Conclusion

The present chapter, in some sense, contains most of the main theses of our project, the theses which are to be clarified as we progress through the phases of our investigation. However, we should emphasize some results of this chapter, which should be used as bases for the rest of the investigation. Then here is some concluding remarks:

1. Inner dialogue is possible, and, considering its close relation to expression in its purity, i.e. before that indication comes to stage, we should regard it as a prototype in studies on language and dialogue. Of course, outer dialogue, in some strong sense, has something more; and to concentrate on inner dialogue is by no means enough to study dialogue in general. However, we assert that, considering inner dialogue, we can

reach profound theses about the constituents of dialogue and argumentation in general.

2. Inner dialogue, as primordial, is wordless. It is, only in this sense, of crucial importance in philosophical studies. We should not confuse inner dialogue in this sense with the stream of words occurring in mind, either emerged as a consequence of inner, wordless dialogue or as a psychological phenomenon. This second case is not of genuine value in so far as we intend to investigate the essence of expression, language and argumentation.

3. A very important point in respect with the constitution of the meaning is that expression in its originality is different from indication. Expression can be present in the states wherein indication could not work. Inner dialogue is of importance, for it is the archetype of such a situation. In accordance with this, we see and reinterpret this idea of Husserl that indication and sign are not relevant to thinking and reasoning as such.

In complete agreement with this is Brouwer's thought in that mathematics is languageless (noticing that Brouwer's conception of language is that of public language, namely a system of signs). Husserl and Brouwer say that in the efforts pertaining to knowledge, including our studies on meaning, dialogue and logic in so far as supposed to increase our knowledge, we should beware of confusion between the sign and what for which the sign takes place. Accordingly, we should not attribute the relations belonging to signs qua signs to the realm of the signified, e.g. the realm of meanings in the case of expression.

Chapter 2

Meaning and the Unintuitive

2.1 Introduction, Statement of the Problem

In section 1.5, we dealt with the constitution of meaning. However, the kind of discussion which was presented there is not enough to study the theory of meaning in general. In fact, in the previous chapter we focused only on the *primordial*; and in this way we linked the expression with the intuition. Nevertheless, we should take into account the cases in which meaning is at use without having a genuine experience of the object meant. Accordingly, we should consider the place of the meaning as detached from the intuition, and investigate the acts functioning in the realm of meanings as such, namely regardless to the (possible) intuitions behind them.

The case of *symbolic thought*, that is, thought that works by means of mere symbols and without referring to intuition, is one of the controversial problems in this respect. For those who do not believe in the relevance of intuition in thought[1], symbolic thought

---
[1]For example, for proponents of formalism, structuralism and behaviorism,

is not an exceptional case. But for proponents of phenomenology and also of intuitionistic approach to logic and mathematics, this problem is a very significant one.

In the present work we advance in the clarification of construction of meanings, having an idea of the constitution of primordial meanings in mind. Here is an outline of this investigation.

First of all, we should make clear the difference between primordial and non-primordial. It is not the case that every meaning is primordial. By act of expression in its primordial sense, we discover, and get access to, the realm of "logos". However, once we get access to the logos, we can constitute and be conscious of those meanings which are not results of primordial acts of expressing. We should distinguish, while speaking of acts of transcendental ego, between primordial acts or products from one side and ordinary or natural ones from the other side—when the latter come to stage as inhabitants in those realms which are already constituted by the primordial acts of transcendental constitution. The distinction between primordial and natural is similar to that of construction and habitation (or utilization) or that of practice and habit: The latter would not exist if the former does not, however the latter has an extension of its own and it could not be reduced to a function of the former. We will explain more about this distinction and its phenomenological place.

Related to the mentioned distinction is the idea of *sedimentation* which is introduced by Husserl to explain a special phenomenon about the intellectual productions. We will briefly explain this idea and its relevance to the theory of meaning.

---

and some other viewpoints in Analytic philosophy, intuition is nothing special, it is nothing but a psychological issue reducible to belief or feeling.

For phenomenology, intuition is a faculty of the transcendental ego and should be kept apart from the psychological; this latter, as Husserl shows in the second investigation in the LI, could not provide a ground for meaning. This is the case also for intuitionism. In this school, intuition, as taken as the ground for some sort of knowledge, is not treated as psychological. (see (van Atten, 2004))

2.1. INTRODUCTION, STATEMENT OF THE PROBLEM

Husserl in the sixth investigation of his LI speaks about the difference between *intuitive intentions* and *signitive intentions*. If we want to develop a theory of meaning on the base of the theory of intentionality, we must follow this distinction which, provisionally, would lead in a two-layered account of meaning constitution. The distinction between intuitive and signitive has some precedent in Husserl's early work and also he returns to it in his manuscripts for a revision of the 6th investigation. I will explain these themes, then depart to analyze the constitution of the meaning in the absence of intuition and on the basis of some signitive intentions.

In the next phase, considering these remarks alongside with certain novel ideas, we will step forward to study the "pure grammar", namely the structure of realm of meanings. An important issue in this respect and about the synthesis of meanings is the problem of fulfilling of categorial intentions (like those behind "and", "or", "all" and so on). A considerable part of this chapter will be dedicated to illustrations of the peculiarities of categprial intentions as well as categorial intuition .

Having discussed the constitution of complex meaning and the issues around categorial synthesis, in the next section I will study the statue of different types of meaning. Since for most of the existing theories of meaning, proposition is the basic unit of meaningfulness so that other forms of expression are supposed to obtain their meaning on the basis of their role in respect with the proposition, I will analyze the case of proposition in more details than other cases. Proposition is of course of prominent importance since it is the bearer of truth or falsity; and it is the ideal content of judgment which is the constituent of any form of argumentation.

At the final step of this chapter, before going to the discussions on dialogical logic, I will present a summary of the phenomenological account of pure logic. I will also discuss Brouwer's argument against logicism, which is in accordance with Husserl's viewpoint, as stated in his 1908 paper.

2.2 A Place for the Unintuitive

2.2.1 The difference between the primordial and the non-primordial

In the sphere of experiences and acts of the ego, Husserl distinguishes between two general cases:

> [We have] *in the first place* the distinction between positional experiences in which what is posited is *given primordially*, and those in which it is *not* given in this mode;[2]

This distinction between the primordial and the non-primordial, or between the original and the ordinary[3], plays a significant role in understanding the various aspects of human being's lived experiences. First of all, we have this distinction in respect to expression and meaning, though "we find similar distinction in all act-spheres" (Husserl, 1969, p. 380). In the case of expression, we can understand, say, a sentence in two different manners: as a declaration of an intuition or perception, or as a mere claim. One may believe that any proposition is asserted in order to reveal a (positional) experience.

---

[2]"Fürs Erste den Unterschied zwischen positionalen Erlebnissen, in denen das Gesetzte zu originärer Gegebenheit kommt, und solchen, in denen es nicht zu solcher Gegebenheit kommt." (Husserl, 1976, §136)

[3]I use the word "ordinary" instead of words like "indirect" or "derived", because I think the function of the mentioned distinction is to distinguish primordial experiences from ordinary ones. Only afterward, we see that the other experiences are derived from, or mediated through, the primordial ones. Indeed, we have a lot of such experiences without that they *prima facie* appear as indirect, derived or founded, rather, we have ordinary experiences that we, in phenomenological attitude, seek for grounds for them, grounds in some special experiences which are named positional or primordial. I do not think that to use the terms like "indirect" to refer to non-primordial acts is wrong, but I think that the terms like "ordinary" more suites to recall them as far as we speak about a phenomenological, descriptive distinction.

2.2. A PLACE FOR THE UNINTUITIVE

Nevertheless, we can realize that if such an account is taken for serious in this simple way, a significant part of intellectual effort would be neglected. In fact, we should acknowledge the phenomenological observation that in most cases there is no direct relation between expression and perception (or intuition in general). We still can say that such an expression is not *originary*, in the sense that it depends on some other positional expressions as its constituents.

Nevertheless, the distinction between original and ordinary is not *substantial* for the meanings or expressions themselves. The property of being primordial is not primarily attributed to meanings or expressions but to acts behind them, and thus the distinction between primordial and non-primordial by no means divide the realm of meanings into two in-itself, namely substantially, different spheres. However, this distinction is a very important one, for without considering it one may lead to the position to see all acts, expressions and meanings either as original or as ordinary. The possibility of non-primordial expression is the basis for the possibility that a primordial expression, or an expression which is presented in order to manifest an intuition, can be understood, or its meaning can be grasped, by a person who himself or herself has no such an intuition.

A significant place in which we can see the principal role of non-primordiality is communication and its functioning in culture and civilization. Husserl introduces the notion of *sedimentation* to analyze the role of concepts and judgments as detached from the intuition, especially in the context of a tradition, after they were once originated.

> propositions, like other cultural structures , appear on the scene in the form of tradition; they claim, so to speak, to be sedimentation of truth-senses [Wharheitssinnes] that can be made originally self-evident; whereas it is by no means necessary that they actually have such a

sense.[4] [The translation is mainly that of Carr (Derrida, 1989, p. 170) except that I use the word "sense", instead of "meaning", for the translation of the German word "Sinn".]

As Husserl puts it, "sedimentation" appears as a ground for the possibility of habits, culture and tradition in general. The idea behind this claim is that although a concept or a judgment can be, at first, formed intuitively, in the everyday use of it or in regard to its function in a course of a progressive activity, e.g. in science, we do not recall that concept or judgment in a fully intuitive manner; rather we treat it, during the mentioned activities, as *potentially* intuitable, but actually sedimented. Therefore, for instance, in progress of reasoning our intuitive concentration turns toward the vary innovative moment, to the, so to speak, ultimate stage of the course. We have not to, and actually do not, recall all previous steps and all required items as intuitive or self-evident. However, we are confident that we can go back to each of those items and make them present to us in their being intuitive or self-evident; namely, we can *animate* them.[5] This confidence make us able to go forward in, say, reasoning and focus our glance to what we want to reach without spending time and power to actually animating every step. The phenomenon just described has been named "sedimentation" by Husserl. Husserl therewith asserts:

> We can also say now that history is from the start nothing other than the vital movement of the coexistence and

---

[4] " Nun treten aber Sätze wie sonstige Kulturgebilde als solche, als Tradition auf; sie erheben sozusagen den Anspruch, Sedimentierungen eines ursprünglich evident zu machenden Wahrheitssinnes zu sein, während sie doch,..., keineswegs einen solchen haben müssen." (Husserl, 1954, p. 377)

[5] This confidence may be wrong; and it is possible that in a case of a frustration or inconsistence we go back and revise what we had taken for granted as intuitable. However, it is not an ordinary move in the progress of a tradition, including the development of a scientific theory, and it involves a kind of "revolution".

2.2. A PLACE FOR THE UNINTUITIVE

> the interweaving of original formations and sedimentations of senses (Sinnbildung und Sinnsedimentierung).[6] (Derrida, 1989, p. 174)

This holds not only for intersubjective sphere but also for a person alone, for his own history and the progress in his own course.

Here we should mention, in accordance with Derrida in his introduction to Husserl's *Origin of Geometry*, that:

> This prescription [the necessity of sedimentation] in turn is sometimes valued as the condition of historicity and the progressive advent of reason, sometimes devalued as what makes origins and accumulated sense become dormant. It truly is a threatening value. (Derrida, 1989, p. 36 f.)

The fact that the ego can live non-primordial acts (in its various sorts) provides it with major practical advantages, but carelessly progressing in them may result in loosing the original intuitions. For example if in the progress of reasoning we always take a concept in its sedimented mode and do not pay attention to its ability to be awaken, then we may loose its hitherto unnoticed intellectual consequences and if there have been a mistake in the formation of that concept we would not recognize it and this mistake will repeatedly appear in every reasoning based on this concept. The mere confidence of the animatability of a concept or judgment should not substitute actual animating when needed. We may also diagnose the phenomena that sometimes the role of sedimentation turns out to be totally transparent; and, as a result, intuition seems to be practically irrelevant. That is, in the usual practice of a science,

---

[6] "Wir können nun auch sagen : Geschichte ist von vornherein nichts anderes als die lebendige Bewegung des Miteinander und Ineinander von ursprünglicher Sinnbildung und Sinnsedimentierung."(Husserl, 1954, p. 380)

I replaced, as well, the word sense in lieu of the word meaning in Carr's translation.

once an intuition of a notion is achieved, a scientist may only pay attention to its relations in under-developing formulas and see no need to fully clarify this notion's intuitive status in every practice, and this seeming autonomy of a part of formal and theoretical expansion may be, by some observers, harmfully generalized to the whole practice of a science, as if there is no intuition is required at all.

However, on the other hand, emphasizing on the importance of intuition and primordiality should not make us negligent to the fact that the ability of having non-primordial acts is itself a capability[7] which is primordially present to the ego.

Therefore, a specific phenomenological investigation is needed to analyze the non-primordial acts or, in other words, non-intuitive intentions. Husserl did this in part of the sixth investigation of *LI*. After introducing some kinds of non-intuitive intentions (for example imaginative ones), he concentrated on signitive intention which is of a prominent importance in regard to the meaning construction. In the theory of signitive intention we find a phenomenological explanation of Leibnizian idea of "blind thought" or of what Husserl himself used to call "symbolic representation". In the rest of this section we intend to elucidate this theory.

2.2.2 The theory of signitive intention

> The formation of every mathematical concept which unfolds itself in a chain of definitions reveals the possibility of *fulfilment-chains built member upon member out of signitive intentions*. We clarify the concept $(5^3)^4$ by having recourse to the definitory presentation: Number which arises when one forms the product $5^3.5^3.5^3.5^3$. If we wish to clarify this latter concept, we must go back to

---

[7]we use this word here in the same sense that Husserl has introduced and used the German word "Vermöglichkeit" (Husserl, 1939, §8) to indicate "Möglichkeit als Vermögen", that is, possibility as capacity [for the ego].

2.2. A PLACE FOR THE UNINTUITIVE

the sense of 5^3, i.e. to the formation 5·5·5. Going back further, we should have to clarify 5 through the definitory chain $5 = 4 + 1, 4 = 3 + 1, 3 = 2 + 1, 2 = 1 + 1$. After each step we should have to make a substitution in the preceding complex expression or thought, and, were this proceeding indefinitely repeatable — it is certainly so *in itself*, just as it is certainly not so *for us* — we should at last come to the completely explicated sum of ones of which we should say: 'This is the number $(5^3)^4$"itself"'. It is plain that an act of fulfilment not only corresponded to this final result, but to each individual step leading from one expression of this number, to the expression next in order, which clarified it and enriched its content. In this manner each ordinary decimal number points to a possible chain of fulfilments, whose links are one less in number than the number of their component units, so that chains of indefinitely many numbers are possible *a priori*.[8] (Husserl, 2001b, p. 723)

---

[8]"Jede in einer Definitionskette sich entfaltende mathematische Begriffsbildung zeigt uns die Möglichkeit von *Erfüllungsketten, die sich Glied für Glied aus signitiven Inten-tionen aufbauen*. Wir machen uns den Begriff $(5^3)^4$ klar durch Rückgang auf die definitorische Vorstellung: "Zahl, welche entsteht, wenn man das Produkt $5^3.5^3.5^3.5^3$ bildet". Wollen wir diese letztere Vorstellung wieder klar machen, so müssen wir auf den Sinn von 5^3 zurückgehen, also auf die Bildung 5.5.5. Noch weiter zurückgehend, hätten wir dann 5 durch die Definitionskette 5 = 4+1, 4 = 3+1, 3 = 2 +1, 2 = 1 +1 zu erklären. Nach jedem Schritt hätten wir aber die Substitution in den zuletzt gebildeten komplexen Ausdruck, bzw. Gedanken, zu vollziehen, und wäre dieser Gedanke immer wieder herstellbar (an sich ist er es gewiß, obschon ebenso gewiß nicht für uns), so kämen wir schließlich auf die vollständig explizierte Summe von Einern, von der es hieße: das ist die Zahl $(5^3)^4$ "selbst". Offenbar entspräche nicht nur dem Endresultat, sondern schon jedem einzelnen Schritte, welcher von einem Ausdruck dieser Zahl zu dem nächst aufklärenden und ihn inhaltlich bereichernden überleitete, wirklich ein Akt der Erfüllung. In dieser Art ist übrigens auch jede schlichte dekadische Zahl eine Anweisung auf eine mögliche Erfüllungskette, deren Gliederzahl durch die um 1 ver-minderte Zahl ihrer Einheiten bestimmt

In the phrase brought above, Husserl describes an instance of cases in which we deal with the signitive intention. Husserl introduces the notion of signitive intentions in §14 of sixth investigation and this becomes an important theme for arguments up to the part §27 of that investigation. While some intentions can have a direct act of fulfillment corresponding to them, some intentions can be fulfilled through a *chain of fulfillments*, namely by means of mediated fulfillments; and thus these intentions are directed toward those objects which have not been presented intuitively.

To explain the above analysis, consider the judgment $2 = 1 + 1$. Such a judgment could be intuitive; namely we can have a direct intuition which fulfills our intention toward $1+1$ be equal to 2. Both 1 and 2 are able to be grasped by the intuition—here intuition should not be confused with mental image. The ego may have intuitive intentions toward 1 and toward 2. And the truth $2 = 1 + 1$ can be as well the object of an intuitive intention. However in order to grasp the truth of $5+7=12$, we need to analyze the arguments in order to reach intuitive elements. Also, in order to grasp what $(5^3)^4$ is, we do not refer to direct intuition; our intention toward $(5^3)^4$ may be fulfilled through a chain of fulfillments. So we have an intention here, and a sequence of intentions, which are not intuitive; they only signify their objects without intuitively giving it. The ego should follow the chain of this signitive acts in order to reach a fulfillment, or experience an unfulfillablity.

Accordingly, if I say "the door is open", when I express this as a result of my experience of the state of affairs, the intention behind such an attitude toward the state of affairs is intuitive. However it is possible that I claim that the door is open without actually seeing the door. In this latter case I have an intention which is not based on an intuition of its object but turns toward it in an indirect manner; and thus it is possible that there would be no

---

ist, so daß derartige Ketten von unbegrenzt vielen Gliedern *a priori* möglich sind." (Husserl, 1968, p. 69)

2.2. A PLACE FOR THE UNINTUITIVE 99

such an object at all. I may infer, in this or that way, that the door is open. Therefore my intention is fulfilled through a chain of fulfillments, hence a signitive fulfillment. Also it is possible that I blindly claim that the door is open and it turns out to be closed, so my signitive intention remains unfulfilled. While an intuitive intention is fulfilled intuitively, a signtive intention may be fulfilled, signitively or intuitively[9] or not fulfilled.

The difference between signitive and intuitive intentions can be described in that:

> A signitive intention merely points to its object, an intuitive intention gives it 'presence', in the pregnant sense of the word, it imports something of the fullness of the object itself.[10] (Husserl, 2001b, p. 728)

Intentionality is one of the central themes in phenomenology. Every conscious act is intentional, and on the other hand every objectivity is constituted by an intention. But not every intention is intuitive. Namely it is not the case that every intention would be based on an intuitive relation with an objectivity. This is already clear if we take into account that the relation of intention to its object is not a "real" relation. The ego intends something; it is possible that this thing itself evokes the intention, but intention as such does not depend on the existence of the intended nor on intuition of it. So it is also possible that the ego intends something in the absence of its corresponding intuition. Every object is constituted by an intention. According to the mentioned observation, not every

---

[9]That means a sigitive intention, e.g. when I claim it is raining out there, may be fulfilled signitively, e.g. by means of an inference, or intuitively, e.g. by means of seeing the event. So a same object may be intended by, and fulfills, now a signitive intention and then an intuitive intention. This is a very important phenomena which is to be explained in the following.

[10]Die signitive Intention weist bloss auf den Gegenstand hin, die Intuitive macht ihn im prägnanten Sinne vorstelling, sie bringt etwas von der Fälle der Gegenstandes selbst. (Husserl, 1968, §21)

object must be constituted as real, namely as intuited or intuitable. Then in the case of objectifying intentions, it is possible that an intention brings forth its object as truly existent, in the case of intuitive intention, or brings forth its object in a peculiar way which can be called sigiftive representation.

> If we bear in mind the fact that the same (e.g. sensuous) content can at one time carry a meaning, and at another time an intuition—denoting in one case and picturing in the other—we are led to widen the notion of a representative content, and to distinguish between *contents which represent signitively* (signitive representatives) and *contents which represent intuitively* (intuitive representatives).[11] (Husserl, 2001b, p. 739)

The line of thoughts explained above is in accordance with Husserl's previous investigations in *Philosophy of Arithmetic*, where Husserl asserts that we have direct comprehension of small numbers like one, two, three etc.; and others are represented through arithmetical operations. One can recognize the resemblance between the general distinction of signitive and intuitive intentions, made in *Logical Investigations*, and the distinction of *symbolic* and *authentic* representation of numbers, made in *Philosophy of Arithmetic*. Husserl have there explicitly spoken about "the fundamental fact that all number representations that we possess, beyond the first few in the number series, are *symbolic*" (Husserl, 2003, p. 200). Husserl have even claimed:

> If we had authentic [eigentliche] representations of all numbers, as we do of the first ones in the series, then

---

[11]Beachten wir, daß derselbe (z. B. sinnliche) Inhalt einmal als Träger einer Signifikation, das andere Mal als Träger einer Intuition dienen kann (hindeutend - abbildend), so liegt es nahe, den Begriff des repräsentierenden Inhalts zu erweitern und zwischen signitiv und intuitiv repräsentierenden Inhalten (oder kurzweg: signi-20 tiven und intuitiven Repräsentanten) zu unterscheiden.(Husserl, 1968, p. 89)

2.2. A PLACE FOR THE UNINTUITIVE

there would be no arithmetic, for it would then be completely superfluous.[12](Husserl, 2003, p. 201)

It means, in the latter terminology, that arithmetic is to deal with [a sort of] signitive intentions; if all intentions toward numbers were intuitive then there would be no place for arithmetic.

We think that Husserl's thoughts on this subject, from the time of *Philosophy of Arithmetic* to that of *Logical Investigations* and also to that of the *Ideas* and revisions in LI in 1913, can be interpreted as advanced in a somehow same line. About the place of the theory of signitive intention in the framework of thoughts represented in *Ideas1*, especially considering the theory of noetic-noematic correlation, we will say more in the following section. Nevertheless, about the importance of thoughts explained in *PA*, just in contrary with the usual account that assumes that the central thoughts of that work have been abandoned by Husserl and that work suffers from, thereafter refuted, psychologism and thus has nothing to do with genuine phenomenology, we are in agreement with those scholars who assert that there is no radical change in transition from *Philosophy of Arithmetic* to *Logical Investigations*, though there is an exaggeration in a self-criticism made by Husserl in order to avoid then-more-probable misunderstandings of his philosophical project. Moreover, Husserl himself in his *Formal and Transcendental Logic* asserts that the most of arguments in the *PA* are still valid. I am in agreement with Farber's viewpoint when he remarks "... this work was never wholly abandoned. On the contrary it provides strong evidence of the continuity of Husserl's development." See for further discussions (Farber, 2006, p. 59).

---

[12]Hätten wir von allen Zahlen eigentliche Vorstellungen wie von den ersten in der Zahlenreihe, dann gäbe es keine Arithmetik, denn sie wäre vollkommen überflüssig.(Husserl, 1970, p. 191)

2.2.3 Signitive intention and meaning

Non-intuitive intentions have various kinds. In the sixth investigation of *LI* Husserl mainly deals with this theme in relation to expression and meaning. In order to clarify the point that it should not be assumed that signitive intention functions only in relation to meaning, Husserl mentions some roles of "signitive intentions beyond the limits of the meaning-function" in the paragraph 15 of that investigation. In the revision of this part, which was written in 1913 but remained unpublished until 2002, Husserl develops this discussion and dedicates a whole section to it (§§15–19 (Husserl, 2002, pp. 85–99)).

Since signitive intention plays an important role in the theory of meaning, a brief look at the different types of the non-intuitive intentions in order to have a distinct idea of signitive intention as related to the meaning would be helpful. Hence, in this section I discuss Husserl's analysis of the issue in his manuscript for the revision of the sixth investigation.

First of all, in the function of wording, there is a signitive intention. When I say, or write, "It is raining in Paris", you understand me, because you share the intention behind such an expression. However my intention toward the alleged state of affairs may be intuitive or not, the hearer's intention is in the first place non-intuitive. You go through my words to the meaning and an objectivity is represented but not in an intuitive manner. So in general, in communication we have the signitive intentionality at work. The intention that the verbal expression evokes in the hearer is a signitive intention.

Nevertheless, there are other types of non-intuitive intentions in which verbal expression has no place. An important instance, is the case of wordless recognition, when you recognize something *as* something (Husserl, 2001b, p. 716). In this case the recognition is possible because there was a concept of the objectivity which is now perceived. The unification of fulfillment which takes place in the recognition is the correspondence of a current intuition with an

2.2. A PLACE FOR THE UNINTUITIVE

intention which was empty before. Assume that I take a look at the table and see a book on it. I notice that I see *a book* even without recalling the word "book" in my mind. So, there was a non-intuitive intention, which makes the recognition possible, which is not related to the word, though there is still a meaning-function here.[13]

Moreover, in the perception of an external objectivity there are a sequence of interrelated intentions at work. For example when I listen to a melody, a kind of expectation of the following notes and a kind of remaining atmosphere of the precedent ones makes a melody what it is. Therefore not only the intuition of the present note and not only the intention directed toward it but the now-not-present moments are functioning in the constitution of the object as such. There are intentions, which although are empty, are indispensable in the representation of a transcendent object, like a melody or a spatial thing.

Husserl says:

> Only in this way the thing can appear: that for it one side "actually appears" (territory of the actual intuition) and other sides, in a more or less undetermined manner, "non-actually" appear (territory of apprehension).[14] (Husserl, 2002, p. 89)

Accordingly, Husserl states that:

> Here, *every intuitive intention is surrounded by a halo, by an areola, of empty intentions* and it is with them one in essence.[15] (Husserl, 2002, p. 90)

---

[13]This part of Husserl analysis remains almost the same in the revision (Husserl, 2002, p. 85).

[14]Nur dadurch kann das Ding erscheinen, dass von ihm beschränkt eine Seite "eigentlich erscheint" (Bereich der eigentlichen Intuition) und anderes in mehr oder minder unbestimmter Weise "uneigentlich" erscheint (Bereich der Apprehension).

[15][H]ier *jede intuitive Intention sozusagen mit einem Hof, einem Strahlenkranz von leeren Intentionen umgeben* und mit ihm wesenseins ist.

Therefore, non-intuitive intention functions in both conceptual and non-conceptual experiences and in both meaning-understanding and sensual perception. Thus Husserl, in the mentioned revision, prefer to use empty intention instead of signitive intention, in order to stress this broadness, and reserve the latter term for those empty intentions which are in relation to meaning.[16]

In respect with the empty intention, Husserl speaks of empty modification (*Leermodifikation*) of the intuitive representation.[17] Every non-intuitive intention is in connection to one or more intuitive intentions; this connection may be based on a association, e.g. in the case of a spatial perception, or a non-associative *excitement* (*Erregung*). So, every empty intention is brought about as a modification of intuitive intentions in this or that way; and in all of them the intuitive representation of the intention is emptied. In the case of perception of a material thing we have a sensual, thus intuitive, representation, which is brought about as a fulfillment of an intuitive intention. For example I see the front side of the book on the table. However my awareness of that very book contains

---

[16]There are also different kinds of empty intentions in regard to meaning. In the mentioned text, Husserl uses the terms "signitive" and "significative" somehow as synonym—"signifikativen oder auch kurzweg signitiven" as preferred to the term "symbolic" (Husserl, 2002, p. 39). However in a close look, he prefers to use the term "significative" while speaking of the intention concerning wording and expression (Husserl, 2002, p. 74), and the term "signitive" in other cases concerning meaning. In other manuscripts on the issue, published in (Husserl, 2005), he makes the distinction explicit—the difference of which has been discussed and explained by Melle (Melle, 1998). However, In the present study I will use only the term signitive since our interest is directed toward meaning in general and we are not concentrate on the intentions behind wording in contrast to other intentions toward meaning.

[17]In the texts written for a revision of the sixth investigation Husserl uses the idea of modification and modification-to-empty which is in accordance with the analysis of neutralization, as a vital theme of transcendental phenomenology, in the book *Ideas I* that Husserl had just worked on. For a detailed study of the role of this theme in phenomenological theory of consciousness see (Brainard, 2002).

2.2. A PLACE FOR THE UNINTUITIVE 105

intentions to other sides and other features of the book which are not presently perceived. These intentions are not fulfilled, they are empty, but they are essentially connected to the intuition of the perceived side: the signitive representation of not-yet-perceived features are at work as modification of, and in this case as associated to, the intuitive representation. However, the case of non-associative connection between empty intention and full intuition is much more important for the theory of meaning and also conceptual knowledge.

Assume that I see that the snow is white. So I have an intuition of a state of affairs, hence an intuitive intention toward an objectivity. However, I can have an intention, and express it in the sentence "the snow is white" when I have no actual perception. The intention may be modified to empty and constituted as a signitive intention which may function in the absence of the actual intuition. The phenomenon is the same as described in the beginning of the chapter as the relation between primordial and non-primordial. Moreover, in this case there is a very important point. Husserl says that, in general, knowledge in the proper sense is obtained thanks to this empty modification. First, we have a *static union of intuition*; namely in the case of actual perception, the intuitive intention and its fulfillment coincide. However, by modification to empty, an empty intention, i.e. a signitive intention is brought about, so its distinction with its fulfillment is thematized. Then not only the object itself but the moment of fulfillment is experienced. Thus in the experience of the fulfillment of a signitive intention, in this transition from empty to the full, in this *dynamic union of intuition*, the ego experiences *truth*.

Accordingly, Husserl considers modification to empty as the general consciousness-modification (*Bewusstseinsmodifikation*) (Husserl, 2002, p. 96).

A signitive intention, may be constituted as a modification of an intuitive intention or as a categorial synthesis of other signtive representations which finally must rest on some intuitions—in a modification relation and in a non-associative connection. About categorial synthesis we will discuss in the rest of this chapter.

The distinction between signitive intention and intuitive intention allows us to speak about the possibility of construction of some meanings which are not direct products of primordial expressing; that is the possibility that not every meaning is a result of such an act of expressing which was supposed to raise content of experience to the realm of logos and give it an ideal meaning. Therefore, the mentioned possibility is the possibility of being conscious of some meanings which do not come to stage primarily in experience of the objects (if any) corresponding to them.

This is the peculiar point: Logos is not only the realm of ideal entities which are brought about as meanings emerged in the course of experience, but also it has its inner structure which makes it possible to construct other ideal entities, namely other meanings, in that realm.

2.3 Categorial Synthesis

2.3.1 Signifying feature of meaning

As we saw above the empty modification through which the ego deliberately turns an intuitive representation to a signitive representation is one of the essential features of the consciousness. This act is the same as that which was analyzed in chapter 1 under the title of primordial expression. Here the content of intuition is raised to the realm of the ideal. Here what is obtained is the ideally signitive representation of an objects which is capable to be expressed as such. The meaning, thus, "signifies" an objectivity. However, "signifies" here should be understood in the manner as described above as related to signitive intention and as essentially different from "indication". Indication is based on association: x indicates y for the agent A on the basis of a mental, i.e. psychological, association between x

2.3. CATEGORIAL SYNTHESIS

and y in A's mind. However, signitive intention, and thus signitive representation or the meaning which signifies its object, has nothing to do with psychology; it is grounded on a peculiar ability of the transcendental ego, namely the ability of modification-to-empty and thus primordial expression.

The above point can be better understood if we benefit from Peirce's distinction between three kinds of signs: *Icon*, *Index* and *Symbol*. Roughly speaking, icon functions as indicator by means of a similarity to the object, it shares something of the object and has a experimental connection to it. Index indicates by means of a real connection. But symbol is based on what Peirce calls *Thirdness* and considers it as irreducible to physical, causal relations, nor to mental perceptions; symbol stands for something based on an ideal relation which has its own ideal nomology. Nonetheless, Husserl does not use the word "symbol". He used to employ this term in his *Philosophy of Arithmetic*, but he preferred not to use this term in his further investigation perhaps because of the misunderstandings which the word symbol might cause. Usually the word "symbol" is used when a real, predetermined object is supposed to refer to another real, predetermined object. (Just like for example the color green is a symbol for peace.) However, this is exactly what Husserl wants to avoid: Not the signitive intention nor the signitive representation are real objects, accordingly the meaning of an expression through which the signification to an object is supposed to take place is not a real object itself rather it is essentially ideal. And also the signified object is not necessarily predetermined: in most cases the object is determined through the very act of signification. We know that a sigitive intention may work as an objectifying act; and thus a wide range of objects of experience owe their determination to signitive acts. These considerations made the term "signitive representation" more preferable for Husserl than the term "symbolic representation" which he had used before.[18]

---

[18]The fact that Peirce used the term "symbol" does not mean that he was

Another very important reason to leave the term "symbol" is that this word is related to the written symbols or to the entities of a formal system. In Husserl's earlier viewpoint, symbols and operations among them were responsible for transition to the non-intuitive. Such operation might be understood in a mere formal sense. However Husserl radically refutes his earlier operational view on ideal entities and explain the subject on the basis of intentionality.[19] Since Husserl rejects the idea of operation, puts aside the term symbol as well. In his proper method, which later finds the name "transcendental phenomenology", Husserl speaks about signitive representation and *synthesis* among them, which both depend on the spontaneous activity of the ego.

Categorial synthesis is performed on meanings or on signitive representations not on intuitive representations of the objectivities. About this difference I will discuss later on in this chapter. In the present part I aim to first roughly sketch out how the signitive nature of meanings are relevant to the act of synthesis in order to reach a new meaning which "signifies" its object.. And then I will mention the differences of this theory with the formalist accounts on ideal entities as applied in logic and meaning theory.

In order to explain the signifying feature of meaning, we should first pay attention to the differentiation between primordial and non-primordial meanings. A primordial meaning, take it to be the meaning of a primordial expression as a consequence of an intuition, entering in a synthetic act with other meanings, can partially function in construction of another meaning which is not primordial. However, since this latter meaning has its own identity and can find its own primordial expression, we should admit its

---

not aware of the mentioned considerations. But as his main focus was on semiotics he was trying to make precise his own terminology about the all relevant concepts; and of course one should not interpret Peirce's technical terms in a way they are used in every day speech.

[19]Husserl himself describes his disappointment of the continuation of *PA* on the basis of the mentioned revision (Dodd, 2012).

2.3. CATEGORIAL SYNTHESIS

separate existence. Therefore, it can be said that those former meanings *signify* this latter, for neither the formers dissolve as "building blocks" in the latter, nor the latter has only a fictitious identity. Consider, for example, the meaning of "night". This can be a primordial meaning, i.e. the result of an expression taken place within an experience. It is so for the meaning of "star". Now the combination "starless night" is meaningful. It has a meaning that is signitive in first place. It means that this latter meaning is quite understandable even if we have no perception of a starless night. We have no intuition of a starless night, but we can understand this meaning, and our understanding has to ultimately refer to some intuition(s), e.g. intuitions of night and star. Nevertheless, it is in principle possible that one finds a proper intuition of an object or a state of affairs corresponding to the meaning of starless night. It is possible that we have an experience, and a primordial expression, of this situation. This assures us that that kind of meaning of which we speak here, namely signitive meaning, is not a fictitious meaning or inaccurately considered as meaning.

Accordingly, the suitable explanation here is that the meanings of "night" and "star" and the dependent meaning of depriving (about dependent meanings we will study in the next section) altogether, through a synthetic act, *signify* the meaning of "starless night". So, the intention behind this latter meaning is a *signitive* one, though this very phenomenon may be also subject of an intuitive intention.

2.3.2 Differences of the viewpoint defended here from any kind of formalism

Now, we acknowledge that the ego can construct a meaning on the basis of other meanings, while in a certain respect they all are in the same level (they all belong to the realm of logos and this relation of construction is not a relation of reduction). Would not this be a kind of permission to formalism and structuralism— or another stance which define concepts not out of intuition or by referring to objects but based on their interrelation and their functioning in the communicative context — to return?

We think that there lies a truth in these viewpoints and also we think that truth that formalism or structuralism intend to assert is properly expressed in this phenomenological theory. Moreover, phenomenology clearly specify its basic disagreements with any kind of strict formalism:

1. In the mentioned viewpoints, the structural relation is attributed to symbols or signs (while sign is taken to be nothing but its place in this structural relation). But, for phenomenology, the relation in question is the relation in the realm of meanings and between those are *essentially* meanings. This relation essentially holds in the ideal realm of logos.

2. It is not merely internal relations that constitute logos and meanings, rather structural relations ultimately are based on those meanings constituted by intuitive intentions—or in other words, emerged by means of primordial expressions. Whereas for formalism the relation of elements is self-sufficient, this is never acceptable for us.

It is worthwhile to emphasize again that signitive meanings can not be in general "reduced" to the already constituted intuitive meanings. For, here, even a constructed meaning may find its own intuitive intention. Therefore, the reduction does not hold, in the sense that the intuition of an object, once objectified through a signitive intention, can not be reduced to the intuitions of the constructive factors of its signitive intention.

For example, in the region of numbers, suppose that we have intuition of number 1 and intuition of the function successor. Then, we have intuitive intentions of these two; or, in other words, we have access to their meanings in the realm of logos by means of primordial expressions. On the basis of these meanings a signitive

2.3. CATEGORIAL SYNTHESIS

intention can be synthesized which construct the meaning of 2. However, we could, and still can, have an intuition of 2 itself, namely regardless to its appearance in the relation suc(1)=2. Therefore, although meanings of all natural numbers can be constructed from the intuitive meanings of 1 and successor, they can not be reduced to them. And it is not true that, say, 2 is nothing but suc(1). After having intuition of 2, namely after having an intuitive intention toward an object toward which we once had a signitive intention, we find that 2 is something in its own right which also satisfies the relation x=suc(1).

Yet, it is possible that we obtain no intuition of the object toward which we have a signitive intention. I myself can give the number i or the 4th dimension as example. However, I think that the possibility of getting intuition of them is still open, (unless we can *a priorily* prove that their inner structure is contradictory, i.e. any hypothetical object having such a meaning would be self-refuting[20]).

2.3.3 The interplay between construction and intuition

> If one finally asks how one and the same content (in the sense of 'same matter') can at times be 'taken up' in the manner of an intuitive, and at times in the manner of a signitive representative, in what the differing nature of these interpretative forms consists, I can give them no further answer. We are facing a difference that cannot be

---

[20]For example, "square circle" is a signitive meaning but we can not have an intuition of its object, for as an object it would be self-refuting, in the sense that it immediately destroys itself. Yet it is a genuine meaning and emerged on the basis of some underlying intuitions, e.g. that of circle and square; and exactly on the basis of its very construction out of intuitive constituents we can *a priorily* prove that it is self-contradictory.

phenomenologically reduced.[21] (Husserl, 2001b, p. 742)

In this important passage, Husserl remarks that it is a phenomenological principle that a meaning which has became available through signitive intention can be turned out to be subject of an intuitive intention without any change in its identity in the realm of logos. This is an expression of the fact that we have previously encountered: the distinction between primordial and non-primordial is not an absolute one; this does not divide the sphere of meanings into two separate domains. We should always notice the high importance of this feature: we can primordially get access to something toward which we had already a non-primordial access; or, to put it in other words, we can have intuition of an object which has already been constituted through signitive intentions.

Consider the example of starless night. The ego can construct a meaning on the basis of its intuitions of star and night and deprivation. The meaning emerging by means of this construction can find its own intuitive intention by perceiving the object corresponding to it, i.e. by perceiving a starless night. This is of course a possibility and a constructed meaning may never find an intuition corresponding to it and it still remains an authentic meaning.

However, if a constructed meaning turned out to be intuitive, this would be an expansion in the abilities of the ego, in the possibility of its further constructions.

Suppose that we have an intuitive access to the meaning of "mass". It means that we had experience(s) bestowing the noematic content "mass"; and after the act of expression we have reached a noematic nucleus, i.e. an ideal meaning in the realm of logos,

---

[21] Fragt man nun schließlich, was es macht, dass derselbe Inhalt im Sinne derselben Materie einmal in der Weise des intuitiven, das andere Mal in der eines signitiven Repräsentanten aufgefasst werden kann, oder worin die verschiedene Eigenart der Auffassungsform besteht, so vermag ich darauf eine weiterführende Antwort nicht zu geben. Es handelt sich wohl um einen phänomenologisch irreduktibeln Unterschied. (Husserl, 1968, §26)

2.3. CATEGORIAL SYNTHESIS

pertained to this objective feature. The noema of "mass" can be enriched in different manners, that is, it can include different layers of senses. No matter by which of these senses has one experienced mass, one can grasp the ideal meaning of that. This experience can be included in the experiences of, say, *heaviness, resistance* or *accumulation*. In each one, if the ego does the act of expression, then it will have the same *meaning* of mass—same meaning but different senses. After this moment the noema can be gradually filled in continuity, every time the ego lives an experience in which it can improve its intuition of mass.

Now on the basis of the ideal meaning of mass (which is itself presented on the basis of the intuition of mass) and the ideal meanings of, say, velocity and multiplying, the ego can construct, by a signitive intention, the meaning "$m * v^2$" and name the assumed object objectified by this meaning "energy". This signitive intention lacks a full noema, it has only an ideal meaning which has not supervened on its own noematic stratum but has been signified by other noematic nuclei. However, we can assume a situation in which the ego have experience of energy itself. And it is plausible to assume that the intuition of energy goes beyond to those of mass and velocity, in the sense that energy be present where there is neither mass nor velocity, in the sense of their primitive intuitions.

Meanings are ideal. Therefore, in regard with ideal objects we should distinguish between object itself and its meaning which is ideal as well. For example the object 2 (if any) is different from the ideal meaning of "2". The ideal meaning of 2, whether or not we believe in the existence of the object 2, truly exists. Whether there is an object corresponding to this meaning or not is up to intuition (in this case, categorial intuition). Of course, in regard to our philosophical stance, in accordance with the insights declared within phenomenology and also intituitionism, we acknowledge that there *is* an intuition of 2; then there exists object 2, and this object, regarding to the content of the intuition pertained to it, is ideal—

namely not real nor mental/psychological.

The issue of signitive intention when working with ideal objects may turn out to be a little bit confusing. For a signitive intention introduces a meaning out of some other meanings, and all of these happen in the ideal realm of logos, and the so introduced meaning may correspond to an object which is itself ideal and more likely to be confused with its meaning. However, to work with ideal objects is more interesting if our investigations concern with logic and mathematics. Hence, we think that it would be useful to repeat the example brought above this time in regard with ideal entities like 2 and 3 instead of material entities like mass and energy. This will help us better to understand the dialectic between construction and intuition and its place in the practice of eidetic sciences by focusing on the case of mathematics.

Suppose that we have an intuitive access to the meaning of "2". It means that we had experience(s) resulting in emerging a noema pertained to the object or objective feature "2"; and after the act of expression we reached a noematic nucleus, i.e. an ideal meaning in the realm of logos, pertained to this objective feature. The noema of "2" can be enriched in different manners, that is, it can include different layers of senses. No matter by which of these senses has one experienced twoity, one can grasp the ideal meaning of that. This experience can be included in the experiences of, say, *opposition*, *succession* or *discernment*. In each one, if the ego does the act of expression, then it will have the same meaning of 2. After this moment the noema can be gradually filled in a continuity, since the ego lives various experiences in which it can improve its intuition of 2.

Now on the basis of the ideal meaning of 2 (which is itself presented on the base of the intuition of 2) and the ideal meaning of, say, successor, the ego can construct, by a signitive intention, the meaning suc(2) and name the object so intended "3". This signitive intention lacks a full noema, it has only an ideal meaning

2.3. CATEGORIAL SYNTHESIS

which has not supervened on its own noematic stratum but has been signified by other noematic nuclei. However, we can assume a situation in which the ego have an experience of 3 itself, e.g. in the experiences of "mediation" or of "synthesis". In this case the ego can immediately recognize 3 without perceiving it through mediated fulfillments.

The fact that the ego constructs the meaning of a mathematical object does not conflict with the viewpoint that the ego construct the object itself as well. However, these are two different things. If, for an entity, to be constructed by the ego were the only way of existence, then it will be intuitable only after construction. In this case the ego could have intuitive intention toward it. But the ego can have also signitive intention toward an object which is not yet truly existent and only the *meaning* of it has been constructed. For example we can construct a meaning "square circle" or the propositional meaning "3=2-1", but neither of these are intuitive, it means we can not construct those ideal objects (a proof object in the latter case) themselves.

In each case, construction is based on some primordial intuitions. However, in some cases an intuition can come to stage after construction, and by so coming it makes further constructions possible. Then we should notice that in the act of definition, the status of the defined can go further than that is involved in the definition. In the revision of the sixth investigation Husserl says:

> So, properly speaking, the meaning of the defining term in not the meaning of the defined.[22] (Husserl, 2002, p. 231)

As far as a complex meaning, though ontically produced by the ego, corresponds an objectivity of its own, it cannot be reduced to its constructing meanings. So, the relation between the defined and

---

[22] Also sicher ist, eigentlich gesprochen, die Bedeutung der definierenden Rede nicht die Bedeutung der definierten.

the defining is not unidirectional, as if the former depends on the latter, but there is an interplay that becomes possible thanks to the role of possibility of intuition. Then, for example as Husserl says:

> The representation or meaning "circle" is, however, first of all a meaning in itself, but which in turn interacts with a second, namely the defining representation, in the experience of fulfilling coverage.[23] (Husserl, 2002, p. 232)

About this interplay Husserl is very explicit in the mentioned text. While in the first edition of the *LI* he had spoken in the manner that one might think that, e.g, the meaning of 5 is nothing but $1 + 1 + 1 + 1 + 1$(Husserl, 2001b, p. 723), here he stresses:

> Isn't, one may say, 4 a sign for 3+1, and 3 for 2+1 and so on? But, this is something quite inauthentic and secondary. 4 is name for the number that is obtained when I make the sum of the number $3 + 1$, 3 again is the number which, etc. So 4 is not simply a name for a concerned united group, it is a name for a number...
>
> 4 is a word whose meaning lies not in "the number which is so and so formed from units", but the object of thought "...".[24](Husserl, 2002, pp. 233–4)

Accordingly, I understand the following figure Husserl brings in

---

[23]Die Vorstellung oder Bedeutung "Kreis" ist aber zunächst eine Bedeutung für sich, die aber ihrerseits mit einer zweiten, der definierenden Vorstellung, in erfüllende Deckung tritt.

[24]Sagt man nicht, 4 ist Zeichen für 3+1, 3 Zeichen für 2+1 etc.? Und doch ist das etwas ganz Uneigentliches und Sekundäres. 4 ist Name für die Zahl, die gewonnen wird, wenn ich die Summe aus der Zahl 3+1 bilde, 3 wieder ist die Zahl, welche etc. Also 4 ist nicht einfach Name der betreffenden Einer-Gruppe, sondern Name für eine Zahl.

4 ist ein Wort, des-sen Bedeutung liegt nicht in "die Zahl, welche so und so gebildet wird aus Einern", sondern im Gegenstand des Gedankens "...".

2.3. CATEGORIAL SYNTHESIS

that manuscript as representing the interplay of construction and intuition, or definition and experience:

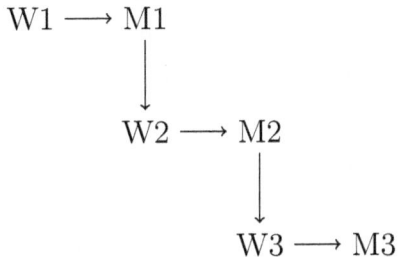

Here W stands for signitive meaning and M for fulfillment. In the case of complicated meaning constructions we experience mediated fulfillment. Suppose that we have intuition of adding (not yet of multiplying), so we have a categorial act to add *two numbers*. Thus in understanding, to use Husserl's example, 3 + 3 + 3 we first have to do 3 + 3. Before that, of course we must conceive of 3 as a number. Having done 3 + 3, we can not go further unless we conceive this latter as a number, namely we have to intuit that there is a number which is equal to 3 + 3. This constructed meaning Husserl shows by W2, while M2 is the experience corresponding to it. Husserl says:

> by M2 I have an intermediate fulfillment, which I have not at W2, I'm only at the gate of the fulfillment. When M2 is a breakpoint, although not an endpoint.[25] (Husserl, 2002, p. 233)

Having conceived 3 + 3 as a number, one can do 3 + 3 + 3 and construct W3, which cues us another fulfillment indicated by M3.

This observation of phenomenology is in accordance with the intuitionist approach to mathematics which says, e.g., a natural

---

[25] bei M2 Ich habe eine intermediäre Erfüllung, die ich nicht bei W2 nicht habe. Ich bin da nur an der Pforte der ersten Erfüllung. Bei M2 ist ein Haltepunkt , obschon kein Endpunkt.

number is not its definition, though the definition can have practical advantages.

Then it is possible to intuit 3, though through 2 + 1, but as an object in its own right. This experience from a meaning-intention to the intuition of its object, is called by Husserl *transitional experience* (*Übergangserlebnis*) (Husserl, 2001b, p. 694). This transitional experience alongside with categorial acts function in the interplay between construction and intuition.

Notice that not every constructed meaning is in the grammatical form of substantivity so that it can be intuited as a simple object. There are of course other categorial forms, such as attribute, proposition, conjunction, etc. which, while to be intuited, are conceived exactly in their categorial form. There are some difficulties about the exact manner of the fulfillment of categorial intuitions which has resulted in some fruitful discussions on the issue (Lohmar, 2006)(van Atten, 2015) (Hopkins, 2011). I will deal with this subject in the next section.

2.4 Categorial Intuition

We have distinguished between intuitive intention and signitive intention. An intuitive intention may be ofa simple form, e.g. in the case of a sensual perception; for instance I see a black book, I have a cognition here of the "black" which is due to an intuitive intention toward an objective feature; the content of intuition here is simple. Nonetheless my intuition of "black book" is an articulated one. This is the case also in the intuition of states of affairs in which the intuitive representation is of the propositional content. Therefore Husserl distinguishes between simple and *categorial* intuitions. Accordingly a signitive representation may be simple or categorial. By means of categorial synthesis a categorial signification may be produced out of some simple signitive representations. In the case of categorial intuition, the categorial form itself is experienced and

2.4. CATEGORIAL INTUITION

intuited. On the other hand categorial synthesis is itself an act of the ego, and thus based on intentions, so that in order such acts to be result-full namely a categorial signification is produced, the mentioned intentions must be fulfilled.

There are some difficulties around the idea of categorial intuition, and there are some debates about what is supposed to fulfill the categorial intention.[26]

How a meaning like "black book" can be expressed? One possibility is the original givenness of an articulated intuition, while the ego experience a black book and expresses it as such. In this case we have first a categorial intuitive intention which begets, after the act of modification-to-empty, a mere signitive intention, whose ideal essence is the meaning "black book". Another possibilty is to synthesize two signtive representations "black" and "book". Assume that the ego has the intuition of black and of book, then modifies these intuitions to signitive representations. Now we have two ideal meanings "black" and "book". Then the ego attribute one to other through a synthetic act, hence the new meaning "black book". How in the first case the categorial form is itself intuited? How in the second case the act of synthesis really results in a new meaning, namely how the synthetic intention is fulfilled? These questions are related to each other and they have caused some discussions around the idea of categorial intention.

As another example, consider the meaning "book and pencil". Again such a meaning may be obtained through an experience in which the categorial form "and" is supposed to be intuited as an objective, though not sensible, moment. Or it can be the result of a categorial synthesis. In this latter case signitive acts must ultimately rest on some intuitions. We may assume that the ego had the intuition of book and of pencil then emptied them and

---

[26]The richest discussions of the questions around categorial intentions are due to Dieter Lohmar (Lohmar, 2002b)(Lohmar, 2006), Robert Sokolowski (Sokolowski, 1981)(Sokolowski, 2000, pp. 88–111)(Sokolowski, 1974, pp. 34–42), Tugenhadt (Tugendhat, 1967, pp. 111-136) and Hopkins (Hopkins, 2011).

then perform the synthesis on the signitive representations. But what about the categorial form "and"? What is the intuition which makes such a feature legitimate to enter into a signitive act? More precisely, what is the ultimate ground of the signitive intention behind "and" and how such an intention is fulfilled? In the previous example, we spoke about the categorial form of attribution and now about the conjunction; the similar problem goes for every categorial form, the problem that how an intention toward a categorial form could have been originally fulfilled so that it can afterwards enter in the signitive acts of categorial synthesis.

Lohmar (Lohmar, 2002b) states that, at least, in the case of the categorial form collection (or set) "the will to perform a synthetic intention is enough to fulfill an intention", namely a categorial intuition is due to will of the Ego. One should notice that "the will" itself does not fulfill the intention but "is enough" to fulfill the intention in the sense that it would be enough if the fulfillment is in principle possible, that is the synthesis would not be against the relevant, eidetic rules. We should distinguish between the will to do something and actually doing it. What Lohmar says is that in the case of intentions of categorial synthesis, the will to perform by the ego is enough for it to actually do the synthesis and thus the intention is fulfilled.

Hopkins (Hopkins, 2011, pp. 421–423) considers such an interpretation unsatisfying and he criticizes the whole idea calling it a sort of psychologism. In his view there is a gap between, say, the "and" or the collection as experienced as an objective feature and the "and" or the collection as emerged in the synthesis. He asks how the form which is produced in the act of the ego is the same as experienced as an objectivity.

> Husserl sheds no light on how the retrospective apprehension transforms the *noetic* "unity" belonging to the "predicatively productive spontaneity" of *collecting* objects together into the *noematic* "unity" of the *collection*

2.4. CATEGORIAL INTUITION

itself. (Hopkins, 2011, p. 415)

Then, Hopkins says if we accept Lohmar's suggestion that a categorial form like collection is produced by an act of the ego then if we attribute this form to an objectivity, this would be nothing but a form of psychologism. (Hopkins, 2011, p. 423)

Van Atten (van Atten, 2015) proposes a new account on the basis of the inner time consciousness arguing that it avoids psychologism. He is concerned with the question that what "representation" would function as the fulfilling character of the act of categorial synthesis. His suggestion is that in certain categorial intentions this representation is based on time consciousness, so that a categorial form, e.g. a collection or a set, is objective from the first place though it is constituted retrospectively.

In this part I wish to develop an interpretation of the status of the categorial intention—which to a certain extent is in the line of Lohmar's original suggestion—and I hope that it would help to clarify some aspects of this significant theme of transcendental phenomenology.

2.4.1 Intuition of the categorial form

Husserl analyses the experience of the articulated objectivity, in which the categorial intuition is at work, in terms of "turning of regard". According to Husserl such an experience involves three stages.[27] To borrow the example from Lohmar, consider an experience grounding the judgment "the door is blue". In the first stage I experience a whole entity, the door with all its perceivable, undifferentiated features. At the second stage I turn my regard to a particular part of the whole perceived stratum, to *blue*. In the third stage I recognize blue as an attribute of the door so that I express my intuition in an articulated form in which "door" possesses the

---

[27]Proper explanations of these stages can be found in the works of Sokolowski (Sokolowski, 1981) and Lohmar (Lohmar, 2002b).

place of subject and "blue" the place of adjective and "is" belongs to the structure of this intuition not to a real moment of the intuited object. However the catogrial form "is" is rooted in the articulated objectivity itself, namely, though in contrast to "blue", "being blue" is not a sensible part of the object of the experience, it is nonetheless objective, and, for example, the difference between table's being blue and its being red is an objective difference. Therefore, it can be said that the categorial forms appertain to the sphere of *formal ontology*.

> *Not in reflection upon judgements, nor even upon fulfilments of judgements, but in the fulfilments of judgements themselves lies the true source of the concepts State of Affairs and Being* (in the copulative sense). Not in these *acts as objects*, but in *the objects of these acts*, do we have the abstractive basis which enables us to realize the concepts in question.
>
> ...
>
> The like holds of all *categorial forms* (or of all *categories*).[28] (Husserl, 2001b, pp. 783-784)

Therefore, for Husserl all categorial forms have the objectivity of their own and they are grasped in the active turning of regard within the act of experience since they are different from the sensuous stratum and need more than a passive perception.

As noticed by Lohmar, the above explanation faces a problem in the case of the categorial form "and", or generally in the case

---

[28]Nicht in der Reflexion auf Urteile oder vielmehr auf Urteilserfüllungen, sondern in den Urteilserfüllungen selbst liegt wahrhaft der Ursprung der Begriffe Sachverhalt und Sein (im Sinne der Kopula); nicht in diesen Akten als Gegenständen, sondern in den Gegenständen dieser Akte finden wir das Abstraktionsfundament ftir die Realisierung der besagten Begriffe;

...

Dasselbe gilt von allen kategorialen Formen bzw. von allen Kategorien. (Husserl, 1968, §44)

2.4. CATEGORIAL INTUITION

of collection. Consider the expression "the table and the window". Table and window can be objects of perception, but what is the objective articulation making them capable to enter in an experience expressible by "the table and the window"? In contrast to, say,"is" it seems that here it is only the free act of the ego which puts two sequences of experiences together without this together-ness corresponding to any objective feature. For this reason, Lohmar considers "and" as an exception; and he says that in the case of collection, the categorial intention is fulfilled by the will of the ego alone. I am going to argue that this is not an exception. The categorial intention, in a certain sense, is fulfilled due to the free will of the ego.

In the following I will argue for this thesis, then I will analyze the constitution of the categorial intention and its different types. I will particularly discuss the constitution of the categorial form "set", and try to illustrate that the phenomenological analysis of the matter, comprehended with all its peculiarities, is free from any kind of psychologism.

In order to develop the proposed thesis, I am going to explore the categorial intuition from another angle than that of the standard explanation mentioned above.

Categorial forms are ideal entities. The ideal, according to Husserl, is objective. Anything objective is object of an intention, according to a fundamental principle of transcendental phenomenology.

Consider the expressions "and", "or", "if" etc., these are dependent meanings. Also we have the categories of substantivity and adjectivity[29] which cover independent meanings. Meanings as ideal entities are themselves objects. Then it can be asked what is the general kind of intention directed toward meanings? or in which kind of acts meanings are intended?

---

[29]"der Substantivität und der Adjektivität"(Husserl, 1974, p. 310).

Concerning the question of fulfillment in the case of the acts involving the intention toward, say, "and", we may begin by the questions what kind of object "and" is, how it is constituted and what role it plays in the intuitions involving it. Is such an object itself intuitive? namely is it possible that the intention toward it be intuitive or it is always a signitive intention? Contemplating about these questions, we will be able to clarify some important aspects of the theory of categorial acts which is in a close relation, as it will be clear, with the theory of *pure grammar* and correlatively that of formal ontology.

Let us begin with the premise that there is an intention; and its object is, for example, the ideal "and" or, say, "set" (which is different from the intention whose essence is of the categorial form corresponding to "set" —and thus can be called categorial intention—but its object is a concrete set). All categorial forms are objects of some intentions. In the following I first try to remove an ambiguity around the expression "object of intention", and then analyze the process in which our objects (the categorial forms) are constituted.

2.4.2 Ambiguity of the expression "object of intention"

When we say that something is an object of a specific intention, two different meanings may be recalled. One is that there is already an object, and an intention is determined as directed toward it, and the other is that by means of a specific intention, once fulfilled, an object is constituted which is thus called the object of that intention.

We should keep in mind that in general neither the fulfillment nor the directedness are prior to, or constitutive of, the intention. Some scholars state that there is always an entity that is responsible for the directedness of the intention prior to the intention itself—they interpret "Noema" as this entity—but such an account is indeed

2.4. CATEGORIAL INTUITION

against the transcendental attitude by reducing the intentionality to a natural relation.[30] However, it is not excluded that in *some* cases an intention is constituted together with its fulfillment—such a kind of intention is called intuitive intention—or also an intention is determined as in some way directed to a predetermined objectivity. In this latter case what is supposed to fulfill the intention is discernible from the objectivity toward which the intention is primarily directed, otherwise the intention would be trivial and it would be pointless to say that the object is constituted by the intentionality for it seems that we have determined the object as such *before* intending toward it, which implies admitting the natural attitude. In order to explain the difference between the objectivity toward which an intention is directed and the objectivity which fulfills the intention I bring an example.

Consider the expression"I hope Mr. and Mrs. Smith are happy". Forgetting the moment of "hope", we have here an intention toward the happiness of Mr. and Mrs. Smith. Such an intention may exist whether or not the happiness is the case. If the happiness holds then the intention would be fulfilled. However the intention is directed toward the life of Mr. and Mrs. Smith, namely this latter situation can considered as the objectivity of the intention (the first meaning). Our intention here is directed toward a predetermined objectivity so that our conscious act (of hope) is *about* something. Such an act is fulfilled by a feature pertaining to this objectivity, namely if such

---

[30] As it can be seen here I emphasize my disagreement with the so-called West Coast interpretation. Of course I can not go to the details of the criticism against this interpretation, however I prefer to be clear from the outset about my opposition in order to stress the point that if in the following I argue in a way that is not in a complete agreement with the East Coast interpretation, namely that of Sokolowski, Cobb-Stevens, Drummond and others, it by no means imply that I prefer the existing alternative interpretation. I strongly reject Fregeian interpretation of the Noema and the corresponding reading of the phenomenological reduction, however I think there are something neglected in the reading of, for instance, Sokoloswki, and I hope that my discussion would help to explore them.

a feature holds it fulfills our intention, so the [fulfilling] object of our intention is happiness (the second meaning). The distinction between these two meanings of the "object of intention" holds also in the case of categorial acts. The general region of objectivity toward which the ego intends in the acts of categorial synthesis is the region of ideal meanings or, as Husserl calls it, the realm of *logos*.

Now consider the strength case in which if I intend that Mr. and Mrs. Smith be happy then they become happy, the happiness holds as it depends only on my intention. The case is similar to this in the act of categorial synthesis.

2.4.3 The realm of logos and the act of categorial synthesis

In the realm of logos, namely in the objective domain of meanings, the intention toward forming a new meaning serves as its own fulfillment.

First of all nothing psychological is here. There is a domain of objectivity, the realm of logos, that its nomological character is so that new objects can be produced by the free will of the ego. For example consider the meaning "table" and the meaning "green", now I have an intention toward them that the latter is attributed to the former, hence the meaning "green table". Or similarly I can produce the meaning "table and window". Notice that although the ego is free in so doing, there are ideal rules governing this realm and it is not the case that every synthesis leads to a meaning.

> Great, however, as this *freedom of categorial union and formation* may be, it still has its *law-governed limits*.[31] (Husserl, 2001b, p. 821)

Thus if we have said that the will to perform a synthetic intention is sufficient for its fulfillment, we had assumed that the intention

---

[31]Aber so groß diese Freiheit kategorialer Einigung und Formung ist, sie hat doch ihre gesetzlichen Schranken. (Husserl, 1968, p. 187)

2.4. CATEGORIAL INTUITION

accorded the ideal rules of pure grammar. That is why, in explaining Lohmar's idea, we mentioned that we have to distinguish between the will and the actual doing, for it is possible that a will does not accord the eidetic rules so it does not result in actual performance although there is no external obstacle. In other words, the fulfillment depends only on the intention of the ego but still it obtains only if it does not violates the rules of the proper realm in which the action is done, namely the rules of the realm of logos or the pure grammar.

Therefore, in two sense the categorial forms can be object of an intention: 1- in the sense that they are supposed as existing and are used in an act of categorial synthesis, and 2- in the sense that they are produced by means of an intention.

We emphasize again that categorial forms are brought about not only by means of categorial synthesis but also through *primordial expression*. Consider the expression "green table". As I described above this can be a meaning produced by categorial synthesis. In this case it is an empty meaning, though both "table" and "green" are *full*, namely there are intuitions behind each. Such an empty meaning ("green table") serves to constitute a signitive intention. Imagine that I say "there is a green table in John's office" without actually seeing it. I can say "green table" in a meaningful way because I have intuitions of "green" and of "table" and I have performed a synthesis of attribution, but I use this expression only in a signitive way. However, imagine that I first experience the state of affairs and then assert the mentioned expression. Here we have the three-stage experience that is mentioned above. Here the intention behind "green table" is intuitive and it is articulated in its essence. So the categorial form of attribution is brought about due to intuition. However such an intuition is not merely passive. It needs an activity of the ego, the activity that arises a mere perception to the realm of the ideal in which the ideal form of attribution emerges as an ideal entity, as a (dependent) meaning.

The meaning of, say, table, is different from the image or any sense that one may or may not have of it. It is important to notice that the categorial synthesis is performed on the meanings not on the intuitive senses nor on the objects themselves. In the case of intuition we may possess a mental sense of the object at work. These mental senses may be in some ways associated to each other. For example the sense that I have of green may be associated to the sense that I have of peace, but such an association has nothing to do with the meanings of these objectivities. So in order better to understand the peculiarity of categorial synthesis we should emphasize its distinction from what Husserl calls *aesthetic synthesis*.

2.4.4 Categorial synthesis versus aesthetic synthesis

In some places of his *Experience and Judgment*, including the paragraphs 43 and 61–63, Husserl analyzes the constitution of "set" (Husserl, 1973, pp. 187–189 and pp. 244–253). The analyses there presuppose the distinction between categorial synthesis and aesthetic (or sensuous) synthesis. Without paying enough attention to this distinction, the preciseness of Husserl's discussion can not be grasped and it even may seem incoherent. Husserl speaks about the mentioned distinction in *Ideas II* (Husserl, 1988, §9). There he declares that the synthesis which is done in a passive manner among perceptual senses, though important in perception of a unity, is incapable of objectifying the experience in a conceptual way; and it should be distinguishes from the categorial synthesis in which the ideal meanings, not the mental senses, enter into a combination so that any object whatsoever can be freely used in such a synthesis while the aesthetic synthesis depends on the material and psychological characters of the objective features at work. Accordingly, he says:

2.4. CATEGORIAL INTUITION

> If we seek to delimit both of these in their peculiarity, one against the other, then we find, as a first distinguishing feature, that the *categorial* synthesis is, *as* a synthesis, a spontaneous act, whereas the *sensuous synthesis, on the contrary, is not*. The synthetic connection is itself, in the first case, a spontaneous doing and performing, a veritable activity; in the second case it is not.[32] (Husserl, 1988, p. 21)

For example consider a tree that I experience in front of me. There is a continuity of the sensuous features, that of greenness and of height for example. The synthesis of these features as constituting a single object is done in a passive way. Here it is based on an objective reality. However it is possible that such a synthesis is done in a hallucination or in a mental connotation. On the other hand, in our example of "green table" I put consciously and actively two features together in a specific form. In contrast to the former case there is no cause or stimulation for such a synthesis. Then the meaning "green table" is a result of the *productive spontaneity*.

Now, let us focus on the case of collection, or set. Having the meanings of "table" and "window", the ego can perform a categorial synthesis and produce the meaning "table and window". Now think about the meaning of the object whatsoever. On the basis of this, by a similar act, the ego may bring about the meaning of "two-member set" and also "three-member set" and so on and also the meaning of "set" regardless to the number of its members. What fulfills the intention toward "set" itself is the spontaneous act of synthesis. However, when we speak about the set "table, window", its fulfillment is involved the fulfillment of the both intentions toward

---

[32]Suchen wir beide in ihrer Eigentümlichkeit gegeneinander abzugrenzen, so finden wir als ein erstes Unterscheidungsmerkmal, daß die kategoriale Synthesis als Synthesis ein spontaner Akt ist, die sinnliche Synthesis dagegen nicht. Die Verknüpfung ist einmal selbst ein spontanes Tun, eineeigene Aktivität, das andere Mal nicht. (Husserl, 1952, p. 19)

"table" and "window". So the constitution of a collection, like "table and window", is done by a free will of ego, but the intuition of it depends on the intuition of the members. Nevertheless in the case of the collections like "three-member set" since the members themselves are ideal, the unity whatsoever, then it is already intuitive, therefore the intention toward forming the meaning "set" is accompanied by having an intuition of it. Then in the case of "set" as such, what provides the categorial intuition is, as Lohmar says, the will to intend it.

Set as an ideal entity should be distinguished from a collection of some given objects. This latter in its turn should be distinguished from both objective bind among the objects and (mental) association among the objective features. The mentioned distinctions are very important in order to grasp the peculiarity of categorial intuition in the case of collection and to keep it distinguished from other possible intuitions at work.

Collection versus association, belonging together and the whole

Imagine that I perceive a table before me and a window over there. Assume that I already have access to the meanings "table" and "window" so that I recognize both objects as unities in their own rights. As it is explained in the case of categorial synthesis, I can form the ideal collection "table and window". In the described situation, the mentioned collection can be experienced as real for I do have intuition of both "table" and "window", for they are actually perceived by me, and it is enough that these intuitions are *brought together* in order to make the collection conceived as real.

On the other hand, consider the case of the perception of the table itself, I see four table legs, the color of green for example and so on; and all of these I perceive in a continuous experience and all as constituting a single object. The synthesis at work here is the sensuous (or aesthetic) synthesis and carried out in the level

2.4. CATEGORIAL INTUITION 131

of passivity. The four table legs and other features here does not consist a collection but *belong together* and conceived in a unity of intuition. The relations of the features are the relations of actuality—in contrast to relations of the members in a set which are not actual. In this sense we see that members in a collection can be freely put together but in the case of an actual whole we need intuitions of the connections.

> [R]elations of actuality *presuppose intuitions resting on actual connection*, intuitions which are called, in *the narrower and proper sense, binding intuitions*. They constitute a unity of intuition, not only of what is *brought together*, but of what *belongs together*—belonging together in the context of a world (or quasi-world) which, on principle, can be made intuitive.[33] (Husserl, 1973, p. 187)

To these two cases, we can add a case in which we have binding intuitions without them resting on actual connections. For example my intuition of peace remember me of the color green and vice versa. Or the number three may have a psychological connection with "plurality" while the number two has that with "conflict". The connections do not pertain to those meanings, and in general are not ideal features. However, they constitute the so called "mentality", which as such is not equal to consciousness. The relation here is called by Husserl *association*. The associated elements, in contrast to the members of a set, are not freely put together, rather there is always a motivation or a stimulation based on reality or psychological facts or social impacts and so on. Also the associated elements,

---

[33]die Wirklichkeitsbeziehungen auf wirklicher Verbindung beruhende Anschauungen voraus, die im engeren und eigentlichen Sinne zusammenhängende Anschauungen genannt werden. Sie bilden eine Einheit der Anschauung des nicht nur Zusammengebrachten, sondern Zusammengehörigen — zusammengehörig im prinzipiell anschaulich zu machenden Zusammenhang einer Welt (bezw. Quasi-Welt). (Husserl, 1939, p. 221)

in contrast to the features of an actual unity, do not necessarily belong together in reality, for they may be under the influence of psychological drives or any aesthetic vision (in the narrower sense of the term). Therefore, and more importantly, in this latter case no intersubjective object is constituted.

A set by itself is intersubjective, for anyone can freely perform the act of synthesis. Since the act here is an act of the transcendental consciousness and is based on ideal, universal entities, it is, as such, different from the mental activities which may differ from person to person due to the issues mentioned above.

If the set is real it means that all members are real objects. For example the set {table, window} is real because the elements are capable to be intuited. But the set {table, centaur} is not real. So far we have distinguished the following four cases: collection or set as such namely as an ideal object, actual collection, actual unity, mere association. An actual collection may be also an actual unity if beside the intuition of members there is a intuition of continuity of perception. For example I perceive before me a collection of thousands trees and I perceive them not just separately but also as a unity, hence there is the object forest. Such an object is called *whole* in which we have both actual collection and actual unity.

> The unified "whole" of the collection becomes objective (theme) in the particular sense if a continuous apprehension [of these objects] one by one and of their totality takes place.[34] (Husserl, 1973, p. 189)

It is important not to confuse the analysis of the constitution of set as such with the constitution of whole. Such a confusion is easy to be committed because in most cases the categorial intuition of a collection holds when a categorial intuition of a whole takes place.

---

[34]Es wird das einheitliche "Ganze" der Kollektion im besonderen Sinne gegenständlich (Thema), wenn hierbei eine durchlaufende Einzelerfassung und Zusammenfassung erfolg. (Husserl, 1939, p. 223)

2.4. CATEGORIAL INTUITION

This is usually the case while beginning with the above mentioned three-stage explanation of the categorial intuition. In that case it is difficult to differentiate the intuition of the whole from the intuition of the collection as such. That was the reason that I preferred to begin from another angle. Now, I hope, we are able to easily see the distinction between the following three which all of them are at work in the customary examples of the categorial intuition of collection:

1. Mere collection, or set, as an ideal entity

2. Actual collection

3. Actual whole

2.4.5 Three possible meanings of "categorial intention"

We are familiar with explanation of the experience involving categorial intuition; the explanation is based on *turning of regard* and analyzes the experience to three stages. We are also familiar with the acts of categorial synthesis. Considering these points, three meanings of *categorial intention* can be distinguished:

1. In the case of categorial intuition, the intuition is of a categorial form. The intuitive intention so fulfilled can be called categorial intention.

2. An intention toward formation of a categorial form, i.e. the intention which is responsible for categorial synthesis, can be called categorial intention.

3. A *signitive* intention of a categorial essence, which can be fulfilled signitively, also can be called categorial intention. (The same objectivity so intended may be also the object of an intuitive intention.)

Recall the distinction between intuitive and signitive intention. In the experiences in which we explain categorial intuition we are already concerned with the intuitive intention. Most of Husserl's discussions in *Experience and Judgment* are so, when he speaks about the *originary self-givenness*. I perceive for example a white wall. Through a three-stage process, which is known for us, I express my experience by asserting "the wall is white". Here I have an intention toward a state of affairs and this intuitive intention here is of a specific categorial form due to the peculiarity of the intuition at work. When I assert that the wall is white, I assert a judgment which is evident, and this judgment, regardless to its evidence, has the ideal form which is the *proposition* "the wall is white".

On the other side, the proposition "the wall is white" can be produced by means of a categorial synthesis, provided that we have the meanings "wall" and "white". Here again we have an intention now toward the proposition itself (not toward a state of affairs). This intention is also intuitive because it is immediately fulfilled by will to form the mentioned proposition. Here we have a specific kind of intention, that its objects (in both sense that I explained above) dwell in the realm of logos. This kind of intention may be called, in a different sense with the former, categorial intention. The intention to produce "set" and other logical and mathematical objects are of this kind.

There is another important issue to deal with. Once a meaning is formed by means of a categorial synthesis, e.g. the proposition "the wall is white", a new intention can be formed on the basis of this meaning, an intention which is toward an objectivity, other than the ideal meaning itself, and so in a *signitive* way. I may have the propositional intention that wall is white, and such an intention may be fulfilled or not, but its objectivity is a possible state of affairs. Before having an ideal unity, I could not have the intention toward its instantiations except in a case of an experience of its

2.4. CATEGORIAL INTUITION

originary givenness. But if I have access to an ideal meaning, I can have the intention toward an instance of it, even without having any intuition of such an instance, thus the mentioned intention would be a signtive one. In this way I have a signitive intention toward, say, square circle or "the wall is white". Even if my intention is fulfilled, it would not be equal with the first case of categorial intention, for the fulfillment can be signitive.

For example, I form a proposition "the wall is white", then I intend a state of affairs to be so. It is possible that in some way I infer that the wall is white, so that my proposition is true, or my propositional intention is fulfilled. But it is fulfilled not intuitively. "Two is greater than one" is a categorial intention which for me is intuitive. But "π is not an algebraic number" is a categorial intention, though fulfilled, is not intuitive. For the first case it is possible to say that the categorial form itself is obtained on the basis of the originary self-givenness of a truth. But for the second case the categorial form has to be obtained through an act of categorial synthesis. However, this is the case for me; it is not excluded that for another ego the truth of the second be intuitive and he or she obtains the categorial form through an ideation over that intuition. Nevertheless, in any case there are categorial intentions which are not intuitive, and their fulfillment needs an analysis different than that of intuitive categorial intention.

2.4.6 The fulfillment of the intention toward the categorial form

I would like to conclude that while dealing with the constitution of the ideal categorail forms themselves, such as "set", to a large extent, we have to be concerned with the second kind of intentions listed above, and to a certain extent we have to deal with the first kind in which the ideal form can be obtained through ideation. It is possible the ideal object is produced by the ego. It is produced in an ontic sense.³⁵ In order for an entity to be objective there is no need that it isn't produced by the ego. This production has nothing to do with the psychologism if we pay enough attention to the distinction between categorial synthesis and sensuous or aesthetic synthesis. Here the fulfillment of a categorial intention is due to free will of the ego and the nomology of logos, which includes *Pure Grammar*.³⁶

In the case of "set", van Atten suggests that its constitution is based on the inner time consciousness. This could be true if we defend the idea that every set is obtained through an original ideation. However, a set can be formed by means of a categorial synthesis in so far as it observes the ideal rules governing its proper realm. If we seek for the roots of *eidetic intuition* of a set, independent of any sensual perception, van Atten's argument is quite convincing. However a set as a categorial form, as based on "and" or "or", can be also produced by the act of categorial synthesis whose fulfillment is different from that of the original ideation. In the case of eidetic intuition we have intuitive representation, which acts as the fulfiller of the categorial intuitive intention. Nevertheless, in the case of fulfillment of the caregorial synthesis, the object of the intention, as I have explained above, is a categorial form itself, so it is not a representation of the object but the object itself which fulfills the intention—of course this object itself as meaning, dependent or independent, is a representation, precisely speaking a signitive representation, of another alleged objectivity. Meaning is ideal and in this sense is ideally immanent in the consciousness, so there is no place for "representation" to fulfill the intention rather the

---

³⁵In *Experience of Judgment*, Husserl, while distinguishing between the "figurations" of an object and the object itself, says that these figuration are product of the ego in the ontic sense while the object itself may not be so(Husserl, 1973, p. 252). By figuration he means different meanings which can be attribute to a same object; and this meanings as "objects of understanding" are ideal and productions of the ego.

³⁶For example, to synthesis "table" and "or" in a form "table or" does not give an ideal meaning.

2.4. CATEGORIAL INTUITION 137

immanent object itself, if any can be produced, does that.

> Alle formal rationalen Gegenstande sind ideal immanent. Alle materialen rationalen Gegenstande nicht minder. Hierher gehören natürlich alle apriorischen Wahrheiten (Wesens-wahrheiten) und wie wir auch sagen konnen, alle apriorischen Sinne (Bedeutungen), ...(Husserl, 1974, p. 390)
>
> All formal rational objects are ideally immanent. All material rational objects are not so. Here [the former] of course includes all a priori truths and, we can say, all a priori senses (meanings),..

The meanings which are result of categorial synthesis, are totally immanent so they are "adequately" present in the consciousness and they themselves, not any representation of them, fulfill the intentions toward them. But yet, although they are adequately present, and in this sense they are immanent, they are ideal and universal and possess their own ontic place. To this latter point I will come back at the concluding section of the chapter.

Van Atten's idea suggests to see similarities between phenomenology and intuitionism. Regardless to the case of "set" and the proposed relevance of time consciousness to its constitution, I think the basic intellectual affinity between the two is true and very important in order better to understand both schools. For transcendental phenomenology, as for intuitionism, ideal entities are constructions by the ego, or the creating subject, while such constructions are spontaneous and at the same time has their own *law-governed limits*.[37]

Although for the case of categorial intuition the idea of "turning of regard" is very important, we should also pay attention to the case of fulfillment of the signitive categorial intention, in which "turning of

---

[37]For arguments on the basic affinity of phenomenology and intuitionism see van Atten (van Atten, 2010).

regard" is not enough to explain; otherwise some misunderstandings of the whole idea will arise. Husserl in *Experience and Judgment* is concerned, as it is clear from the subtitle, with the genealogical investigations, so he mostly concentrated on the intuitive intentions. But this should not be interpreted as if Husserl posits that the intuitive intention is enough to explain all situations in which we have a categorial form. If we take into account the act of categorial synthesis which can be done in the absence of the intuitions of the meanings at work and also the act of signitive fulfillment of a categorial intention, I think that it would be clear that Husserl's idea of categorial intention is free from any kind of psychologism. But I think some problems will arise if we overlook these latters and take Husserl's genealogical explanations as if they are supposed to work in all relevant situations.

For example Hopkins criticizes Husserl by saying that:

> More precisely, the issue of whether symbolic acts and their intuitive fulfillment (to use the language of the *Logical Investigations*) are involved in the *preconstitution* (to use the language of *Experience and Judgment*) of the collection as an object accomplished in the process of collecting, or in the retrospective grasping of it as an objective categorial form, is passed over in silence. (Hopkins, 2011, p. 423)

But in fact it should be silent, because the intuitive fulfillment of symbolic acts are involved neither in the preconstituition of the collection (or other objectivities of a specific catagorial form) nor in the retrospective grasping. Preconstituition and the retrospective grasping are questions of genealogy and are related to the originary self-givenness, while the passage from a symbolic act constituting a categorial form to a perception of an objectivity articulated in that very form is done through *signitive* intention. I admit that the question of signitive fulfillment is a very important question in its own right but it is not the theme of *Experience and Judgment* in

2.5. COMPLEX MEANINGS

particular and the absence of of dealing with this questions should not be interpreted as a "gap" in the explanation.

2.5 Complex Meanings, and the Special Case of Proposition

There are various categorial forms: category of object whatsoever, conjunction, attribution and so on. In our previous examples, we mentioned attribution and also collection (on the basis of the dependent meaning "and"). One of the most important categorial forms is that of the proposition. Like any other complex meaning, proposition can be either obtained on the basis of a direct intuition and by means of a modification-to-empty of an intuitive representation, which is in such cases called a "judgment", or produced by means of a categorial synthesis on some other simplest meanings. Certain scholars see proposition as the paradigmatic form of meaning so that other meanings are defined on the bases of their possible role in the proposition. This is of course not the case for phenomenology. However since proposition is of a notable importance in meaning theory specially in logical contexts, for in logic we mainly deal with judgment whose ideal form is proposition, it is quite worthwhile to explain the case of proposition in a detailed way. So in the following, I analyze the status of proposition as a meaning category and then in the next section I will investigate some issues around the relation between proposition and judgment and thus its place in the knowledge theory.

2.5.1 The proposition and the state of affairs

The relation between proposition and perception is so that the proposition pertains to the side of meaning-intention while perception appears in the side of fulfilling acts. A state of affairs intended by a proposition may play a role in this process of fulfillment, namely as the correlate of perception. The intention corresponding to the proposition is of meaning-based essence[38]. It means that an inten-

---

[38]"Bedeutungsmäßigen Wesen" as called by Husserl (Husserl, 1913, pp. 150, 417, 436) which is confusingly translated to "semantic essence" (Husserl, 2001b, pp. 374, 590, 605). Such a translation is confusing because the word semantic is originally refers to the context of linguistics or that of mathematical logic. This term has been used, first in linguistics, by M. Bréal in his 1897 book *La Sémantique*, since the beginning of the twentieth century gradually found its own technical usage which differs from what Husserl means by "Bedeutungsmäßigen" while speaking of intentions. Husserl apparently never used the word semantics, though he extensively speak about syntax which nowadays is considered as the counterpart of semantics. So, to speak of semantics or semantic essence in the context of Husserlian studies may make the discussion seems in such a way that as if Husserl speaks about a formal system or about the character of referring or indicating something independently distinguished. In fact Husserl speaks of constitution of some intentions on the basis of *meaning* as ideal:

> Die Identität "des" Urteils oder "der" Aussage liegt in der identischen Bedeutung, die sich in den mannigfaltigen Einzelakten eben als dieselbe wiederholt und in ihnen durch das bedeutungsmäßige Wesen vertreten ist. (Husserl, 1913, p. 421)
>
> The identity of "the" judgment or of "the" statement lies in the identical meaning, which is repeated as the same in the many individual acts and represented in them by their meaning-based essence.
>
> Dieses Bedeutungsmäßige, ideal gesprochen die Bedeutung, ist beim konkreten Urteilserlebnis der Aktcharakter der urteilenden Setzung (die abstrakte Urteilsqualität) in "attributiver" Verwehung mit dem "Inhalt" (der Urteilsmaterie), wodurch sich die Beziehung auf den "Gegenstand", d. i. den Sachverhalt, vollendet. (Husserl, 1913, p. 436)
>
> This meaning-based essence, ideally speaking the meaning, in the concrete judgment-experience, is the act character of judgmental assertion [propositional attitude] (the abstract judgment-quality) attributively bound up with the "content" (the judgment-matter), which through them together the relation to the "objectivity", i.e. to the state of affairs, is performed.
>
> Therefore, in an actual act of judgment, in which an intention toward a state of affairs functions, we have an ideal meaning, in this case a propositional meaning,

2.5. COMPLEX MEANINGS
141

tion which is directed to a state of affairs through a proposition is based on a meaning, and thus essentially belongs to the sphere of expression acts. The meaning constituting the essence of an intention toward a state of affairs, as such, is of a certain category, namely proposition, in contrast to simple meaning or a kind of dependent meaning and so on.

Then in the case of the relation between proposition and state of affairs, we, first of all, deal with an intention; but notice that this intention may be fulfilled or may not be so; it may even be proved to be unfulfillable, that is the intended state of affairs, the state of affairs to be corresponding the proposition, does not exist.

An intention corresponding to a proposition, just like any other intention, may be intuitive or signitive.

It is intuitive if the intuition of the concerned state of affairs, as state of affairs, namely in the predicative categorial form, is present in the constitution of the very intention toward it. It is worth reminding that from the phenomenological point of view it is admitted that we may intuit objectivities in the predicative or other categorial forms. Intuition is not restricted to that of sensuous. The categorial forms are not necessarily imposed by the ego in each particular experience rather they may be intuited as such. Husserl says:

> *As the sensible object stands to sense-perception so the state of affairs stands to the 'becoming aware' in which it is* (more or less adequately) *given.*
>
> ...
>
> *Not in reflection upon judgements, nor even upon fulfilments of judgements, but in the fulfilments of judgements*

---

which serves in constituting the concrete intention. Therefore, the proposition itself as an ideal entity, different from the act of judgment, is instantiated in the latter by forming the meaning-based essence of the intention functioning in the act.

> *themselves lies the true source of the concepts State of Affairs and Being* (in the copulative sense).
>
> ...
>
> If 'being' is taken to mean predicative being, some state of affairs must be given to us, and this by way of an act which gives it, an analogue of common sensuous intuition.
>
> The like holds of all categorial forms (or of all categories).[39] (Husserl, 2001b, pp. 783-784)

In the intuitive case the state of affairs we perceive is expressed by a proposition without this proposition being made in a second order act depending on the constitution or perception of its parts. Contrariwise, here we may even speak of the act of abstraction—in this case different from the more basic act of ideation—in which a concept is given not based on a proper intuition of it but by abstracting it from the propositions of which this concept is derived as a part. Consider the following example:

Assume that one reaches an intuition so expressed: "*ego cogito*". From the phenomenological point of view, it is plausible—avoiding

---

[39]wie der sinnliche Gegenstand zur sinnlichen Wahrnehmung, so verhält sich der Sachverhalt zu dem ihn (mehr oder minder angemessen) "gebenden" Akt der Gewahrwerdung.

...

Nicht in der Reflexion auf Urteile oder vielmehr auf Urteilserfüllungen, sondern in den Urteilserfüllungen selbst liegt wahrhaft der Ursprung der Begriffe Sachverhalt und Sein (im Sinne der Kopula); nicht in diesen Akten als Gegenständen, sondern in den Gegenständen dieser Akte finden wir das Abstraktionsfundament ftir die Realisierung der besagten Begriffe;

...

Gilt uns Sein als prädikatives Sein, so muß uns also irgendein Sachverhalt gegeben werden und dies natürlich durch einen ihn gebenden Akt—das Analogon der gemeinen sinnlichen Anschauung.

Dasselbe gilt von allen kategorialen Formen bzw. von allen Kategorien. (Husserl, 1968, pp. 140–1)

2.5. COMPLEX MEANINGS

the discussion about the accuracy of this particular example—that the concepts "ego" and "cogito" be consequences of the fundamental intuition of "ego cogito" rather than its independent constituents. It means that this proposition has not been obtained primarily by a categorial synthesis, rather it is a reality (Wirklichkeit) which is in its categorial form intuited and expressed.

However, an intention corresponding to a proposition may also be a *signitive* intention. In this case the meaning of this proposition, the propositional meaning, is a result of a categorial act which has been done on some other meanings—simple or complex meanings which are either constituted by primordial expressions, that is on the basis of intuitive experiences, or are themselves constructed by some other categorial acts. Such an intention, namely a signitive intention of a propositional essence, may or may not find its proper intuition. In the other words, this proposition may corresponds to a state of affairs (which would result in an experience of truth) or not (in this latter case it may also be accompanied by an experience of frustration).

What fulfills a propositional intention is called *evidence*. About evidence I will discuss in the next section (about the relation between evidence and judgment) and in the section 3.3.4 (about the difference between evidence and proof). A piece of evidence which makes a given proposition true may hold in perception, in the case that the objectivity expressed by the proposition be sensuous. It is also possible that the evidence come from the categorial intuition if the proposition expresses an ideal truth.

2.5.2 Propositional truth and falsity

We have said that proposition is a category of meaning which may be constituted either by an empty-modification of an intuition of a state of affairs (which can be, material, mental or ideal) or by means of a categorial synthesis. Such a meaning can serve as the essence of a sort of intentions. Propositional intentions may be intuitive or signitive. In the second case the intention may be afterwards fulfilled or not, whereas in the first case it is already fulfilled. A proposition is called to be *true* when the propositional intention is fulfilled. It is *false* when the ego experiences the unfulfillablity of the propositional intention. However, there is an ambiguity concerning the word "false" since it is sometimes used to say that a propositional intention is not fulfilled, which is different from saying that its unfulfillabilty has been experienced.[40]

Proposition is a category of meaning. It is the ideal content of the (declarative) sentence. However not every sentence is either true or false. Recalling the distinction between reference, sense, significance and meaning, which has been discussed in the section 1.4.1, we can say that every well-formed declarative sentence has a meaning which is a proposition but in order to be true or false it has to have also a significance. Therefore, a proposition by itself, as a meaning, is not necessarily either true or false. It can be so due to its significance, or as Beyer calls it its respective meaning. That is, it should genuinely concern a given, and acquainted, discursive context, and correlatively an ontological region. Even if a proposition were the only proposition of the context or of the ontological region it must be considered in such a respect to be said to be true or false. A not significant proposition, like our example in the mentioned section "Some rational numbers are sorrowful", though meaningful, is neither true nor false.

This important point is in accordance with the principle that every truth is that is experienced. Therefore, in order for a proposition to be capable to be true it should first be connected to a possible experience, that is it must genuinely concern a region of objectivity even if it afterwards turn out to be unexperiencable according to the very nature of that region.

Therefore, from the phenomenological point of view, it is wrong

---

[40]Such an ambiguity exist also around the notion of negation, to which I will return in the last chapter.

to say that every proposition as such is either true or false. Accordingly the meaning of proposition by no means depends on its truth or falsity conditions, for a proposition is meaningful even if there were no such conditions. Indeed the truth or falsity conditions stem from the meaning of a sentence, more precisely they are related to the significant of a sentence which is in its turn posterior to, and partly depends on, the ideal meaning.

Moreover, the meaning of propositional connectives are not obtained by means of their truth-functionality. Such connectives are constituted through various judgment modifications, as explained by Husserl in (Husserl, 1952, §§104–114), and employed in categorial synthesis, so that their meaning explanations should be done on the bases of those themes, namely modification and synthesis. I will return to the meaning explanation of the logical connectives in the last chapter.

2.6 Proposition, Judgment and Belief

Judgment is the unit of knowledge (in the sense of knowing-that) and it is the basic element of any argumentation. Judgment is a doxic act of the ego—the term is also used to the product of such an act. In this sense it seems to have close relation to "belief". The place of judgment in the knowledge and its relation to belief is a significant problem of any theory of knowledge. Since proposition is the ideal form of judgment and we are, in our investigation on the theory of meaning, concerned with the status of proposition, and also since we will deal with the dialogical argumentation whose steps consist of judgments, we should grasp a clear idea of judgment, not confused with any other supposedly related theme. In this section I try to clarify the status of judgment in respect to evidence and knowledge and in contrast to belief.

In what follows, I analyze the concepts in question from the ego's standpoint; that is, the primacy is given to the first-person perspective, as if "believe", "know", "judge" etc. are first attributed to "I" rather than to "he/she/it". This is in contrast with most of the works in the contemporary epistemology which take the third person as the subject of investigation while studying, say, what "know" means. The difference of the approach chosen here with the common one will be clear as we progress in our investigation.

Furthermore I would like to remind the following phenomenological principles:

1. Consciousness has the essential character of intentionality.

2. The transcendental subjectivity should be kept apart from the psychological.

Our first thesis which may be obtained from the above mentioned principles in relation to our subject matter is as follows:

Thesis 1: Knowledge and belief are essentially separate.

This can be considered as a matter of terminology; however, we are not dealing with some arbitrary definitions. If we admit the distinction between the realm of the transcendental subjectivity and that of the psychological, then we should distinguish between what is, due to its nature, psychological—and thus should be analyzed by means of psychological themes such as drive, development, behavior and so on—and what is constituted through a certain type of conscious intention. Regarding the common usage we may reserve the term "belief" for the former and "knowledge" for the latter.

It is somehow a common account to take knowledge as a certain kind of belief, including the claim that equates knowledge with the justified true belief. We assert that knowledge is not justified true belief, but not because the adjectives "true" and "justified" are insufficient to characterize knowledge within beliefs; rather because knowledge is by no means a kind of belief. Thesis 1 states that knowledge and belief are two different genera. In the following, as an

2.6. PROPOSITION, JUDGMENT AND BELIEF

important step toward our main argument, I will explain what this thesis means and how it is in accordance with the aforementioned two premises.

2.6.1 Knowledge and Judgment

First I shall deal with an ambiguity around the word "knowledge". As it is well-known within epistemological studies, we should distinguish between knowing that and other usages of the term "to know".[41]

We are to deal with the "know that (wissen)" in which the object is a state of affairs and, accordingly, the content of knowledge is propositional.

The simplest unity of knowledge (in this meaning of the word) is judgment. I say that "I know that 'the sky is blue'", it means that I judge so and the proposition "the sky is blue" by a judgment turns out to be considered as knowledge; otherwise a proposition as such does not constitute a knowledge.

Now we can see that the first thesis is to say that a judgment is not a belief. Our task turns out to compare belief (as the simplest unity of the system, or web, of belief) with judgment (as the simplest unity of knowledge) and to show their essential difference. After that we will obtain a minimal account of the peculiarities of belief in contrast to knowledge; and this would help us not to fall in category mistakes while trying to explain the relation of judgment (as an element of knowledge) with proposition (as a form of meaning).

---

[41]This is not a properly philosophical issue but rather a linguistic one. It is rather an ambiguity in English; and most of the other most spoken languages are clear in this respect. There is a clear distinction between wissen/kennen (German), savoir/connaitre (French), saber/conocer (Spanish), '-l-m/'-r-f (Arabic), bilmek/tanimak (Turkish), which the object of the former is a state of affairs and that of the latter is a substance (a person or a thing). Then this remark is necessary only because this note is written in English otherwise the genuine difference between to know that p and to know object s would be already obvious.

A criticism may arise here objecting to equating judgment with knowledge: a judgment may be true or false (to put it precisely, right or wrong), but a unity of knowledge is supposed to be true, then knowledge (wissenschaft) cannot be equated with a set of judgments. We shall response that it is true that there are differences between knowledge and judgment but as our analysis will show this does not affect while investigating the differences of these two from one side with belief from the other side.

However, it is not acceptable to explain the distinction between judgment and knowledge by appealing to a naive realism which describes knowledge only from the third-person perspective, as if the truth and falsity is accessible otherwise than through judgment. In order to clarify the basis of the differences between judgment and knowledge, we should notice the following phenomenological observations:

1. There is a distinction between act and product; accordingly we should distinguish between judgment as a primordial act of the transcendental ego and judgment as the product of such an act. While the act of judgment requires peculiarities of its own (see below), a judgment that is constituted as the product of this act turns out to be accessible even in the absence of those peculiarities.

2. The original form of judgment is truth-judgment. However there are forms of modifications which constitute judgments about falsity, possibility, assumption and so on. As Husserl (Husserl, 1969, p. 300) explains truth-judgment, or certainty, is the Urdoxa or the intentional back-reference of all doxic modalities. It is not the case that the genus of judgment "simply splits up into certainty, supposal and so forth, as though the relation could be spinning out a series of co-ordinate species, just, as in the case of sensory quality, color, sound and so forth are co-ordinate species." Husserl calls this a radically false doctrine. Then although what constitute ultimate knowledge are

truth-judgments but other types of judgments, being brought about as modifications of truth-judgment, constitutionally belong to the sphere of knowledge.

Considering the above mentioned remarks we are now able to explicate the phenomenological account of judgment in order to compare it with belief:

Judgment is brought about when a propositional intention of the ego is fulfilled by evidence.

By propositional intention we mean those intentions which their essences involve propositional or predicative nature; for example when I intend that the sky be blue. Notice that there is no need to having this intention before: a propositional intention may arise while the ego apperceives a state of affairs, so the perception of this state of affairs functions as evidence which fulfills the recently emerged propositional intention and thus the ego would be able to perform a judgment. This judgment remains even when there is no more concrete access to evidence, and also I may ascribe judgment to an agent about whom I couldn't say whether has access to evidence. In these latter cases I speak about the judgment as a product rather than as an act.

In the following I am going to explain the essential differences between belief and knowledge. The key criteria are the features of being intentional, conscious and originated in the realm of transcendental rather than psychological which knowledge possesses and belief lacks. In this regard the two aforementioned observations would help to distinguish what is though not original knowledge, but should be analyzed as a modification of it and yet is not a form of belief. For example when I say "I think that p", this is a judgment in a modified form, and this differs from saying that "I believe that p". [42] I will explain more about this in the next section.

---

[42]Even the authors who rightly assert, in contrast with the dominant tendency,

2.6.2 Judgment and belief

Judgment is of propositional content; so seems belief too. Regarding to the content, one may say that to consider knowledge and belief as two essentially different genera is arbitrary. However, we should notice that the content of belief seems propositional in the act of reflection, while a judgment has propositional essence from the outset. The point is that knowledge and judgment are intentional and so depend on a meaning-intention, and the ideal meaning which functions here is proposition. But belief is not intentional in its essence and a proposition is afterward ascribed to it. In order to make this clear, notice that intentionality is not a bare relation of aboutness which can be obtained between two real entities. To say that any belief is about something does not prove that belief is intentional. Knowledge as constituted by intention involves the moment of fulfillment, so to strive toward knowledge is meaningful. The effort toward knowledge is voluntary, while belief is passive. Even if someone speaks of effort of belief or says that belief can be voluntary, this would be based on the presupposition that the object of belief is determined beforehand, then intentionality plays no role in the constitution of this objectivity. While in the course of knowledge, I strive toward what that is undetermined so that my intention could be fulfilled or not. Therefore, intention constitutes a correlation between the ego and the object of knowledge. This correlation is one of the significant features of intentionality. There is no such a thing in the case of belief.

Belief as such does not need to be conscious, while being conscious is essential for knowledge, or judgment. We become aware of our beliefs in the acts of reflection or introspection, while awareness

---

that knowledge does not entail belief, and so want to separate the sphere of knowledge from that of belief (for example (Myers-Schulz and Schwitzgebel, 2013)), have considered "think" and "belief" as equal. This is, I would say, a result of fail in observing one of the important features of knowledge, namely the possibility of modification.

2.6. PROPOSITION, JUDGMENT AND BELIEF   151

of an act of judgment or of a knowledge is not obtained through introspection, for it essentially dwells in the alert consciousness. We are not justified to say that knowledge is different from the belief just because knowledge has something more, for this something more is not an accidental feature rather it is essential so that knowledge is different from belief in its essence, and can not be considered as a sub-genre of this latter.

Suppose that I believe that if I check my emails in Sunday mornings, I would not receive any good news. I state that I believe so, but I don't claim that I know that "if I check ...". In my inner experience I see no ground to judge in favor of the mentioned proposition, but at the same time I believe in that. However, I am not denying any evidence and I am not judging without any ground so that my attitude be irrational; I simply have a belief. I may carry out an introspective study to see that why I believe so, but to have a belief is one thing and to be aware of what guides me to that is another thing; whereas this is not the case for judgment: to perform a judgment is not detached from being aware of what guides me to so judging.

Also, knowledge does not entail belief. One may know that q, but yet acts as if one lacks belief in q, or more commonly, one knows that not-q but yet has belief in q. None of these cases indicate that one is irrational, I am trying to introduce a perspective in which knowledge by itself has nothing to do with belief.

Assume that I am trying to find a way to divide an arbitrary angle to three equal angles using only an unmarked straightedge, and a compass. One tells me: Don't you know that it is impossible? And I say: Yeah, I know, but it is difficult for me to believe it. In this case I know a proof for the impossibility of a certain case but yet I have enough motivations to continue trying. I do nothing wrong in the course of knowledge: I am not denying any evidence and I am not judging without any ground. If I tried to deduce a judgment on the basis of my belief which contracts the reasons which we have in hand, this would be wrong; but in the current case I perform an activity but not a judgment.

Although the original form of judgment is expressed by "I know that", its modifications belong to the sphere of knowledge too; and we can observe that they too are different from belief. In the above example, we saw that I may not say that "I know that if I check ..." because I am not in a position of access to an evidence for this. Nonetheless, to say that "I believe that so and so" and to say that "I think that so and so" are not equal, for in the former I report an actual belief of mine but in the latter I perform a modified judgment which, though not on the basis of evidence, expresses a tendency of seeking for it. The difference between "I know" and "I think" is that the propositional intention of the former is actually fulfilled but that of the latter is modificationally so. Notice that to say that "I think that p" is not merely to indicate having an intention constituted by the proposition p. When I say that "I think that p", I am not merely speaking of an intention of mine, rather I am trying to represent a fact but not with certainty.

Here I give an example to show that "think" is in the side of "know" rather than "believe", and in some occasions belief and thought stand in disagreement.

Suppose that I am playing poker with my friends. Here only I and John continue playing. John bets. After some hesitation I don't call and John wins. He shows his carts: He had made a bluff. I say: "I knew it!". Now let's go back in time: John bets. Assume that one asks me: "Do you know that he is bluffing?". I would say: "No, I don't know, but I think so; yet I don't believe he would be able to make such a risk." (It was according to this belief state that I preferred to give up). Since I once thought he raises with air, now, after revealing the truth, I interpret my thought as knowledge. This is of course not an exact report, but yet it is not totally irrelevant, for now, after getting access to evidence, my modified judgment turns out to be unmodified, a truth-judgment or knowledge; and

2.6. PROPOSITION, JUDGMENT AND BELIEF

since this judgment has been made in the past, I use the verb know in the past tense.

Thinking is an activity in the course of knowledge. The moment "I think" serves in the activity of knowledge, the ultimate object of which is truth. This is different from "I believe" in which I report only a psychological state of mine.

In this table I show the important difference between belief and judgment.

	Belief	Judgment (to know and its modification: think, suppose, guess etc.)
Status	Psychological	Subjective
General field	Mentality	Knowledge
Willfulness	Passive	Spontaneous
Ground	Motivation	Evidence

2.6.3 Encounter with some well-known theses

The phenomenological investigation shows that knowledge is not a species of belief, nor knowledge entails belief. Most of the studies proposing such theses have not been done accurately because they have not payed enough attention to the first-person analyses. However, Peirce, in a well-known paper (Peirce, 1877) on the basis of a first-person analysis tries to equate belief and knowledge.[43] He says:

> We think each one of our beliefs to be true, and indeed it is mere tautology to say so.

However, our examples show that it is usual that I believe that p but I don't consider p as true, nor even as to-be-true. When I say "I know that p", I consider p as true, when I say "I think that p", I consider p as to-be-true, but when I say "I believe that p", I only express my awareness of a state, not an act nor a product, of mine. Peirce's description of belief, and of doubt, really sounds psychological and we would agree with him:

> Belief ... puts us into such a condition that we shall behave in some certain way, when the occasion arises.

However, knowledge by itself does not cause in some tendencies. Knowledge originally belongs to the realm of the transcendental then can be considered by the ego in voluntary acts, but it is not effective regardless of the free will of the ego.

In a recent work Myers-Schulz and Schwitzgebel (Myers-Schulz and Schwitzgebel, 2013) show that the standard trend to consider knowledge as a kind of belief is groundless. But they fail to recognize the genuine distinction between knowledge and belief, because, after showing that knowledge does not essentially entail belief, they propose that the difference is that belief results in tendency and knowledge in capacity. Thus they too analyze knowledge as a psychological matter. However, their proposal lacks enough clarity and has been done very quickly without enough discussion. In any case, we can accept that the belief has relation with the tendency,

---

[43] It should be mentioned that according to Peirce's own viewpoint such an account is not psychologism for he has a special definition of the scope of psychology, as it is more clear in his later works. However from the phenomenological point of view, such an analysis is really to appeal to psychology.

2.6. PROPOSITION, JUDGMENT AND BELIEF

as Peirce declares. But what does it mean to relate knowledge with capacity? We already have accepted the following thesis:

> To have a belief b about the situation S means that if one is to act in respect to the situation S, one would have tendency to act in accordance with b.

Now if we replace "belief" with "knowledge" and "tendency" with "capacity", I doubt that we reach a significant thesis. Capacity is too vague to determine knowledge. It adds nothing to our account of knowledge. If I know that k, it by itself does not extend my capacities to act. knowledge can be a ground in order for the ego to extend his alert consciousness and thus execute a self-realization. But none of these are passive and there is no natural relation here between the ground and the result. Also one may have such capacities without having the knowledge. There is a tight relation between tendency and belief (at least they both belong to the mentality) but the relation between knowledge and capacity is too loose to be considered as a defining feature.

Some try to distinguish knowledge from belief by saying that knowledge is certain and belief is not. However this description is indeed related to the difference between knowledge and opinion, or assumption, namely the difference between knowledge in its original form and its modifications. On the other hand, belief has its spectrum of intensity, and we may see that some of our beliefs are very strong while some others are weaker and less effective. Then to appeal to matters of intensity cannot serve to distinguish knowledge from belief.

Some propose to distinguish between knowledge and belief saying that knowledge is obtained through reasoning or argumentative activity, while belief, or certainty in their terminology, is obtained directly (For example Wittgenstein (Wittgenstein, 1969) or Ortega y Gasset[44]). I shall say that there is a truth in this observation

---
[44] See H. C. Capell and E. L. Niño (Capell and Niño, 1998).

because knowledge needs a ground (which is obtained originally or to be obtained through modification) so that the ground can be observed beside the moment of judgment, while no such thing is necessarily observed or distinguished in the case of belief.

However, this very ground does not necessarily lie in the argumentation; it is not the case that any judgment is obtained through reasoning and thus for any proposition which is marked as true there are another set of propositions which lead to it. This is just one possibility. We may also judge on the basis of a direct intuition, for example when I assert "the sky is blue", the ground of so judging is a state of affairs not some other judgments. If I have an intuition of the corresponding state of affairs, I am right in saying that "I know that the sky is blue". Here, there is no need that I be able to reason for this. This shows that Wittgenstein's analysis to distinguish between knowledge and belief, by saying that knowledge depends on reasoning (Ibid. paragraph 243) is misleading.[45]

Furthermore, The distinction between knowledge and belief would not be eliminated if we take a normative standpoint to belief. We may introduce some norms for beliefs, but those would have no relevance to knowledge, rather they would concern with ethics or search for happiness, very important topics in their own right. I may make a decision to control my beliefs and fit them to my knowledge. This would be a big decision, but this is what I am doing *on* my beliefs, it does not determine the essence of belief. However, in general, to have groundless beliefs is not an epistemic problem; the problem is to take them as knowledge or as science. Now if one devotes himself/herself to knowledge, he/she will have

---

[45]We have said that what grounds knowledge is evidence. Reasoning is in fact a kind of indirect evidence, or, phenomenologically speaking, a chain of signitive fulfillments. This latter kind of evidence is a very important one for the science. It has a tight relation with the notion of *proof*. I will examine the case of proof and its relation to evidence in section 3.3.4. In any case there is no need that every evidence be indirect (as it clearly cannot) and accordingly there is no need that that every knowledge be obtained through reasoning.

2.6. PROPOSITION, JUDGMENT AND BELIEF

to rule out beliefs without evidence.

Therefore, I would disagree with Whiting (Whiting, 2013) who wants to study belief by some normative criteria. The point is that any norm for belief as such is irrelevant to epistemology, yet one can seek for norms about beliefs for other reasons.[46]

2.6.4 Does belief need evidence?

According to the phenomenological viewpoint, knowledge and judgment need evidence, since judgment is rooted in consciousness and can be consciously performed when adequate evidence is present. But there is no reason to consider belief as dependent on evidence. However, we may judge about a belief, considering its content, whether it is right or not. Such an assessment would base on whether an assumptive judgment with that very content would be true or not. In order to do this we need evidence. Now the evidence for the truth of a judgment corresponding a belief may be metaphorically interpreted as the evidence for that belief. That is, we may say that that belief is right because there is a piece of evidence for it.

Our thesis for the source of belief is as follows:

> Thesis 2: For every belief there is a motivation.

This is a thesis, rather than an observation, for it aims to suggest a way to a mutual clarification of belief and motivation; but in the negative way it aims to detach belief from evidence.

---

[46]For example, in this respect I think the principles proposed by Richard Swinburne in the field of religion are remarkable (Principle of Credulity and Principle of Testimony). I am saying that belief as such is not normative, but one may introduce some norms in regard to certain fields of activity. Now for example I accept the principle of Credulity, while admitting that credulity has no place in knowledge. This principle is not supposed to constitute the essence of belief, but one may say that it is good to employ it for the sake of well-being. Such discussion do not primarily belong to knowledge theory. (for Swinburne's argument see (Swinburne, 2004).)

A motivation may be apparent or not (evidence is *eo ipso* apparent). The intensity of a belief depends on the strength of the motivation. This motivation, in most cases, in the final analysis rests on the psychological drives such as will to survive, will to reproduce and so on. Knowledge on the other hand is disinterested in its object. Of course we have mixed cases in everyday life. But philosophy is supposed to find a way to purify knowledge. This is done in phenomenology by radicalizing the very disinterestedness by the method of *epoche*.

As we saw before, it is possible that we know something but yet we have not believed it, so that it does not play an effective role on our behavior, as if that knowledge has not provided enough motivation. However it is possible that a unity of knowledge provides us with a motivation to believe the content of that knowledge. This happens when there is no stronger motivation which prevents us from believing in that. Consider the example of angle trisection. There is a proof for its impossibility, then I have some motivation to believe that it is impossible. But there are other motivations which would lead me to try despite of my knowledge. My final belief is formed according to the dominant motivation. However, I may control myself; I may strengthen one of my motivations, by concentration, meditation, or some other practices. Knowledge in some cases serves as a motivation but it is not the case that knowledge of something always motivates me to believe something, except that I already have a motivation to believe anything what I know. When I believe something I have a motivation to do so in the sense that belief could not be neutral to its object. Whereas knowledge as such is neutral in respect with its object. The fact that the ego's decisiveness in striving toward knowledge is not a neutral attitude is not in conflict with the principle that in any act of knowledge the ego's regard to the objectivity at work should be neutral. In other words belief in any case is rooted in interestedness; therefore knowledge can not be a type of belief, for the genuine knowledge involves

2.6. PROPOSITION, JUDGMENT AND BELIEF

disinterestedness—-though it *may result* in forming some interests for the subject.

Although we distinguished judgment from belief, and said that judgment needs evidence and belief does not, we can still observe that there are judgments without evidence. These are of course inaccurate judgments, but if judgment in its constitution depends on evidence, how unjustified judgments arise?

Notice that this question is not about those judgments which turn out to be wrong, while a piece of evidence happens to be recognized as inadequate. This is not our question, the question is about those judgments which are considered as knowledge while there is no evidence, whether adequate or not, for it. We may observe cases in which one states a judgment (not only a belief) without caring about evidence. Disregarding the matter of correctness, how it is possible that I say that "I know that p", while I have no experience of an evidence to-be-corresponded to it? This, however, happens. In the next section I outline an idea in order to analyze this phenomenon.

2.6.5 Judgment without evidence

As we saw before, judgment has modifications in which evidence is not currently in access. For example, when I say "I think that p", I am aware that there is no adequate evidence for p; however, I am in search for it. The problem arises when one use "know" instead of "think" in the above utterance. Then one of the bases of our problem goes back to the issue of modification. The other I shall discuss in the following.

That the nature of thinking is dialogical has been stressed by certain philosophers. However, there is no enough philosophical investigation elaborating this idea. According to phenomenology, inner dialogue, as primordial, is possible and it does not depend on the natural life so that it would be out of play after *epoche*— as it has been discussed in chapter 1. Then nothing is wrong in saying that a transcendental act, like striving toward knowledge, may be of dialogical nature. In fact this dialogue is possible because modification is possible: *the ego can confront itself by modifying its position in this or that way.*

In striving toward knowledge, I may set forth a dialogue with myself: I propose a thesis, I come up with a judgment (which I already have) which seems to contradict that thesis, I defend by proposing a perspective in which there is no such contradiction, and so on.[47]

This dialogue should not be confused with an arbitrary inner speech: the stream of speeches on our minds that have been heard once or more, have become important for us, are recalled frequently due to unconsciousness affairs, or are recombined involuntarily. However, these two types of dialogues are often mixed. Assume that I am thinking about something, I take different positions, pose judgments and assumptions, and imagine some possibilities. Here due to some psychological indications I may sink in the stream of judgments which are not relevant to my topic. I may find myself defending against some positions which are far from plausible. These are done because of some motivations other than to strive to knowledge. Now if I treat the expression in such an impure inner dialogue as the expressions in a genuine knowledge-oriented one, the problem arises.

In the genuine inner dialogue, I (in each position) do not express groundless judgments, rather I express different modifications consciously, namely thesis as thesis, supposition as supposition, challenge as a challenge and so on. However, in a psychological inner speech my opponent may assert that b without any ground, this would be only an utterance of a belief of mine or a belief that I attribute to someone that I take as my opponent in an imaginative discussion. If I am not aware of the impurity of an inner dialogue I

---

[47]The dialogue concerning logical validities, is a certain kind of knowledge-oriented dialogue, which is codified in the dialogical semantics. About the phenomenological grounds of the dialogical semantics and its contribution to the meaning explanation of the logical constants I will discuss respectively in the 3rd and the 4th chapters.

2.6. PROPOSITION, JUDGMENT AND BELIEF

may take b as a judgment. So I consider b as a judgment without having evidence for it; *it is a diffusion of a belief in the realm of judgments.* I may be careful not to confuse my beliefs with my judgments, but if I be ignorant I find myself having a plenty of judgments without evidence. If one does not care about striving toward truth, it is not strange that one is not concerned with distinguishing belief with judgment. However, if one takes care about the truth, but makes some mistakes of the sort mentioned, one is confronted with judgments without evidence, which are of course in any stage correctable.

Therefore, the ultimate responsible for judgments without evidence are beliefs, but not directly, rather through the inner dialogue.

2.6.6 Summary of the section

I shall briefly conclude that:

1. Belief is formed by motivation, and thus depends on the desires and circumstances in the psycho-social life.

2. Belief has no essential relation to evidence.

3. Evidence is supposed to serve for judgment. But it happens that one has judgments without evidence.

4. The ultimate responsible for judgments without evidence are beliefs, but not necessarily beliefs of one's own. A belief that may diffuse in an inner dialogue may be a dominant belief in the community or of a certain person, which due to its dominance is uttered by a hypothetical interlocutor in a mixed inner dialogue.

The proper subject matter of epistemology should be knowledge not belief. And knowledge and the modified judgments related to it belong to the realm of the transcendental then are constituted by the spontaneous act of the ego. Although belief and knowledge are two different genera, they have significant interactions, a few of which were mentioned in this discussion. More investigations may be done in order to clarify these interactions and their effects in both striving toward knowledge and forming the mentality.

2.7 Conclusion: Husserl's Idea of Pure Logic

Husserl's theory of unintuitive thought is very important for his idea of pure logic as well as for his theory of meaning. In this chapter I tried to explain phenomenological theory of meaning using Husserl's arguments about unintuitive thought. Here I intend to summarize the discussion and also sketch out phenomenological account of logic on the basis of the presented analysis.

As we saw, Husserl's treatment with the unintuitive thought is based on the idea of intentionality. Perhaps it would be a good idea first to outline Husserl's early attitude in order better to understand the characteristics of the intentional theory of the unintuitive. I call the early account, which goes back to the time of working on *Philosophy of Arithmetic* and has been abandoned sometime in 1890s, *semiotic theory of the unintuitive thought*. I call it so because there was a notable stress on the role of signs—and for this reason as it will be explained below, that theory was suffering from a kind of psychologism.

2.7.1 Semiotic theory of unintuitive thought

First of all, it should be mentioned that for Husserl, a semiotic relation is a triadic, and not a dyadic, one. This is also admitted, in a strong way, within the transcendental approach. So instead of

2.7. HUSSERL'S IDEA OF PURE LOGIC

a simple word–reference relation we have:[48]

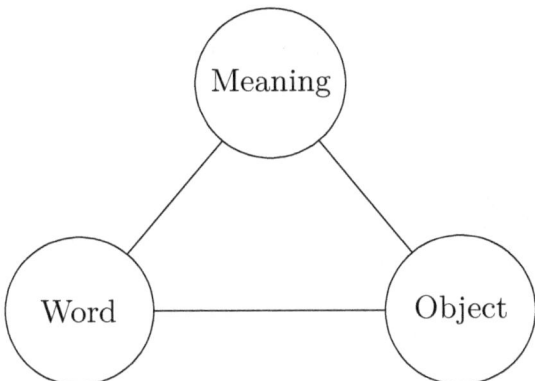

According to this theory there are two kinds of meaning:

1. Authentic, in the case object would be itself given. Sign is assigned to a given object. Meaning is an authentic representation of the object.

2. Inauthentic, in the case object is out of the play. Meaning is introduced as a result of operation among signs.

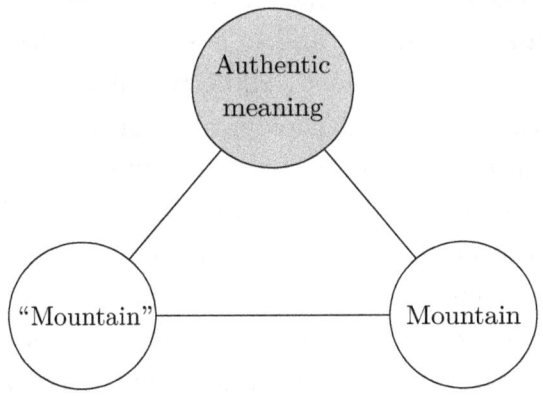

---

[48]In this point Husserl would agree with Peirce and his triadic relation of sign–interpretant–object; though Peirce's theory of semiotic is more elaborate than that of Husserl. Also one can say that Frege's approach admits a triadic relation of signification, in his word–Sinn–Bedeutung distinction. However for Frege we have a tetrad in the case of general concepts: word–Sinn–Bedeutung (here it is the concept)–object (an instantiation of the concept). See Frege's letter to Husserl dated 1891 (Frege and Husserl, 1987, p. 36).

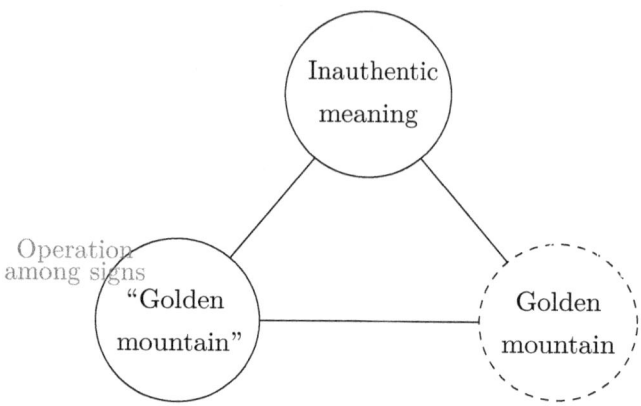

In *PA* Husserl speaks about the great numbers, and thus about any non-intuitive concept, as if they "first come to us through mediation of remote and excessively complicated symbolizations." (Husserl, 2003, p. 249) Here it is the symbols, their physical reality and the apperception of them which is taken to be means to passage to the non-intuitive—and not the mere intentional acts of the consciousness. In this respect, Husserl's viewpoint in *PA* is psychological because he thinks that only a determinate domain of concepts can be intuitive and it is so due to "the limitation of our mental capacities".(Husserl, 2003, p. 249) Therefore Husserl's discussion at that time is psychological as far as he sees symbolism indispensable in respect to the domain of meanings. His discussion is not psychological in the sense that logical or mathematical concepts are mental entities. In contrast to this, he says that there are concepts to which our minds can not have access, unless it does so through the mediation of signs. From Husserl's own later, transcendental point of view, that was psychologicalistic because now he says that the proper way of getting access to concepts as such does not essentially depend on mental issues, as properties of a piece of the natural world; and moreover some ideal entities themselves are produced by the consciousness even if they are unintuitive. In *PA*, he had said, "the narrow confines of consciousness impose insurpassable limitations upon us" (Husserl, 2003, p. 254); but in

2.7. HUSSERL'S IDEA OF PURE LOGIC

the transcendental method the intentionality of the consciousness has no essential dependence on the psychological capacities.

Therefore, this is the short description of early view: there are intuitive objects to which some signs may be assigned and it is done through the authentic representation that we have of the object and such a representation is taken to be the meaning of the sign. There is a "rigorous parallelism" between the systematic of the concepts and the systematic of signs (Husserl, 2003, p. 251). The operation among signs would introduce new signs which would be sign of some non-intuitive objects; and this is carried out on the basis of sense perceptibility of symbolization because the physical appearance of signs has some sensual affect upon us so that leads us to have an "inauthentic representation" of what for which the sign stands. Then in fact, the authentic representation and the inauthentic or symbolic one do really belong to two different genres; so we have two essentially different kinds of meaning. A description of this abandoned view can be find in a text by Husserl written in 1914 (Husserl, 2005, pp. 167–175), translated by James Dodd (Dodd, 2012).

Before going to the phenomenologically appropriate theory, it is good to note two remarks about this early account which are somewhat admitted by the latter theory as well:

1. For a same object, a same objective reference, we may have different meanings (Bedeutung). Here, this is due to different combinations of signs:

    The meaning of "the victor at Jena" is different from that of "the vanquished at Waterloo" though they have a same object.

    Also, 5+7 and 11+1 have different meanings, but refer to a same object.

2. Although on the basis of operational rules among signs we will be able to deal with objects which are not primarily intuitive, it is possible that such an object then be intuited.

2.7.2 Intentional theory of the unintuitive thought

Now this is the intentionality and its peculiarities, rather than sign or its sense perceptibility, that is responsible to passage to the unintuitive, as it has to be in a properly transcendental method. There are two kinds of intentions: intuitive and signitive; and this is the signitive intention which grounds the non-intuitive thought (explained in 2.2.2). Moreover the categorial synthesis (see section 2.3) among signitive representations introduces us to the new signitive representations; so every thing is done in the side of meaning:

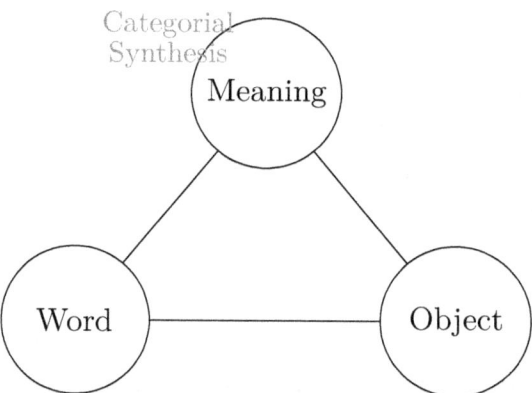

The distinction between the authentic and inauthentic is still at work but the difference is that here primacy is with meaning: first the meaning is constituted then a sign dedicated to it.

However, it would not be right to speak about two different kinds of meaning. An object may be intended both by intuitive intention and signitive intention. Meaning of an expression remains same regardless to whether the objectivity corresponding to this expression is intuited or not.

Here what is relevant is the relation between meaning and objectivity (intuitive intention) or the construction among meanings themselves (signitive intention). Then we may drop the side of sign in our schema:

2.7. HUSSERL'S IDEA OF PURE LOGIC

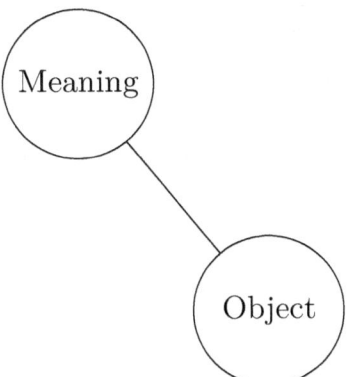

Instead of operation among signs, a synthesis among meanings is performed in order to produce a new meaning, then an arbitrary sign may be assigned to it (or it may follow some combination rules in order better to reflect the characteristics of the synthesis). In any case as far as we deal with the theory of meaning, symbolism is out of play. So, we can show a schema of synthesis as following:

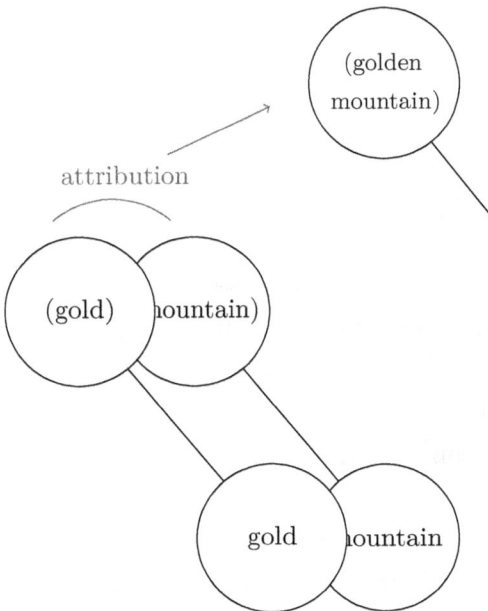

As mentioned above, in some cases it is possible that a meaning which is first constituted signitively finds its intuitive object, in this case it does not loose its identity.

Think about, for instance,' sweet wood' which is fulfilled by cinnamon. The meaning of the former expression remains same because it depends on our intention which remains same whether fulfilled or not.

There is no need to say that we have two kinds of meaning one of which is originated in perception the other in operation on signs. Rather meaning in each case is brought about by intention which itself may be intuitive or signitive.

So Husserl says:

> The meaning is thereby always only one, and it is now empty meaning, then full. (Dodd, 2012, p. 32)

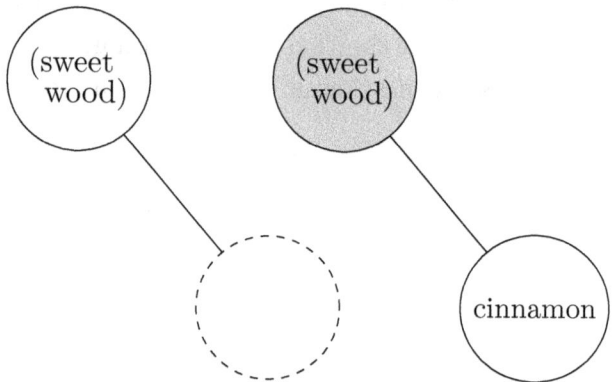

We have said that even in the case of the intuitive representation, the meaning is the result of an empty modification. Therefore, no matter whether or not the object is intuitive, the meaning is taken to be independent of intuition. Then it would be good idea to make a distinction between the fullness of a meaning and the meaning as such. In the occasions in which such a distinction is not important we may use the rems "meaning" and "sense" interchangeably. But, when the distinction is important Husserl prefers to use "meaning" (*Bedeutung*) for that ideal entity which belongs to the realm of logos, and "sense" (*Sinn*) for the fullness which dwells in intuition, as I explained in the section 1.4.1. Then, let's make the distinction:

2.7. HUSSERL'S IDEA OF PURE LOGIC

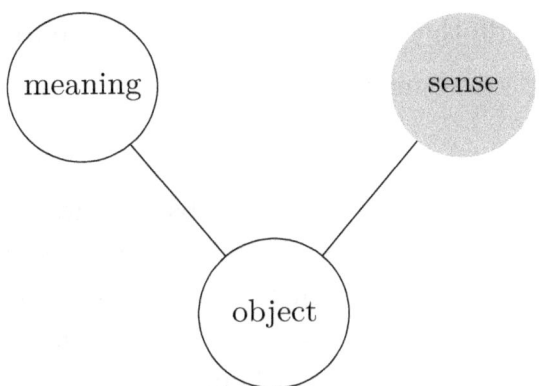

Sense, the mental representation so to speak, has the property of gradualness; it can vary in the course of time, and it may differ from person to person, while the meaning is same and permanent.[49]

For a same object, various layers of sense can be accumulated thanks to numerous experiences. However the meaning would remain identical; and in fact it is the identity of the meaning which grants that a sequence of experiences, despite their different contents, are concerned with a same objectivity and completing a particular sense.

Consider for example the number 2. We may have a sense of separation here or a sense of succession or other senses. The ideal meaning of 2 remains same no matter how one for the first time experienced it. I may intuit it in the moment of succession, but once I express it, I have access to an ideal meaning which transcends my particular experience and provides a nucleus to accumulate and

---

[49]To compare with Frege's terminology, roughly speaking, here object (*Objekt, gegenstand*) corresponds to Frege's *Bedeutung* (usually translated to *reference*), meaning (*bedeutung*) corresponds to what Frege calls *Sinn* (translated to *sense*), and for the notion we specified as sense (*Sinn*) in our schema Frege pays no attention and there is not that much place in his philosophy for the theme of intuition and the how of access to the objects, including ideal ones (our notion of sens somehow corresponds to what Frege called "idea", and this is of course different from the conception of "idea" within phenomenology). Therefore our distinction between Bedeutung and Sinn, in cases that it matters, should not be confused with the Fregean distinction which uses the same terms, namely the sense/reference distinction.

compare my, and others, further relevant experiences.

In this respect we may bring this brilliant quote by Husserl:

> When we raise ourselves to the abstract [realm] we give up the sensuous fullness of intuition, the freshly pulsing life of the individuals, in which we certainly immediately participate. But without abstraction, no concept, without concept, no law, without law no insight into the ground, no theory, no science, and without science no philosophy. The turning away from the green valleys and fields of life full of self-denial to the gray, leathery, sober theory is the only way to satisfy our highest and purest interest in knowledge.[50]

Now we have the new triad meaning–sense–object. In order for an expression to be meaningful it is not necessary that object or sense exist. Meaning is ideal and sense is mental, but object can be real, ideal or mental. Object is the object of experience but also it can be predetermined by its proper meaning then recognized as such. Meaning can be produced by a categorial synthesis but also it can be obtained by means of the empty modification of the intuitive sense. So we have the following relations among our triad:

---

[50]"Die sinnliche Fülle der Anschauung, das frische pulsierende Leben des Individuellen, das unserer unmittelbaren Teilnahme sicher ist, geben wir freilich auf, wenn wir uns zum Abstrakten erheben. Aber ohne Abstraktion kein Begriff, ohne Begriff kein Gesetz, ohne Gesetz keine Einsicht aus dem Grunde, keine Theorie, keine Wissenschaft, und ohne Wissenschaft keine Philosophie. Die entsagungsvolle Abwendung von den grünen Tälern und Gefilden des Lebens zur grauen, ledernen, nüchternen Theorie ist eben das einzige Mittel, um unsere höchsten und reinsten Erkenntnisinteressen zu befördern." (Husserl, 2001a, p. 30)

English translation by Mirja Hartimo (Hartimo, 2012).

2.7. HUSSERL'S IDEA OF PURE LOGIC

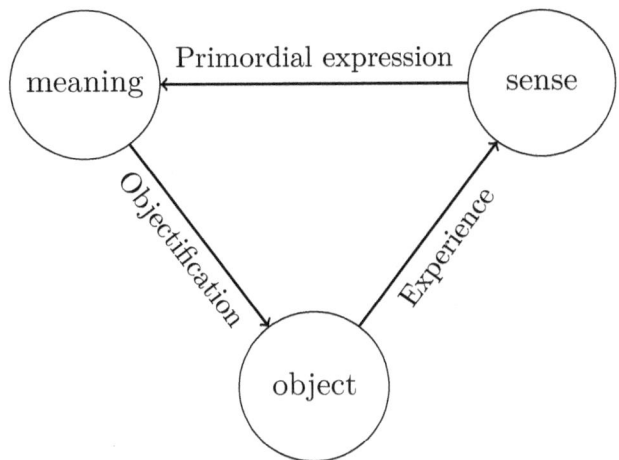

There is another important relation to add. If the object itself would be an idea, it is possible that the ideal meaning is obtained directly from a peculiar experience of the object itself, namely without the mediation of the empty modification. This is called "ideation".

Husserl speaks of ideation and, in that sense, of essence seeing in various places. The following phrase may be illustrative:

> The ideal object, the ideal simple object (I do not mean that composed in categorial synthesis) and also every categorial object that goes to ideals is that in the ideation, in the categorial intuition, is given or to be given completely and firmly.[51]

---

[51] Der ideale Gegenstand, der ideale schlichte Gegenstand (ich meine nicht in kategorialer Synthesis gefasst) und ebenso jeder kategoriale Gegenstand, der auf Ideales geht, ist ein in der Ideation, in der kategorialen Anschauung abgeschlossen, fest, fertig Gegebenes oder Zugebendes. (Husserl, 1986, p. 210)

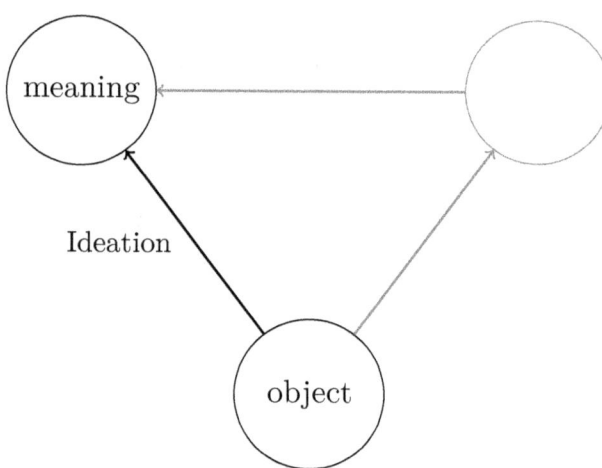

In other words when an object or a moment of it is itself ideal, its entrance to the realm of logos does not need to pass through the medium of sense , rather it may be itself apprehended. This is not to say that sensuous experience and mental representation are absent in the case of ideation, rather that is to say in ideation the ideal object is constituted not by means of modification, for it itself is intuited as such. This grants that there is no circularity in the process of objectification since in each ontological region, in each "manifold", the ego grasps certain basic objective features through ideation which would be the base for further synthesis and thus for further objectification and objective recognitions.

Though sense is very important, the center of logical studies is the ideal meaning and the categorical forms belonging to it, however this does not mean that logic is completely detached from the experience or it is neutral in respect to objectivity. Certain elements of logic may themselves be intuited, for our intuition is by no means restricted to the sensuous ones.

2.7.3 Three levels of formal logic

Proposition is a type of meaning which is of a special importance in knowledge. *Judgment* is the fulfillment of an intention of the

2.7. HUSSERL'S IDEA OF PURE LOGIC

categorial form of proposition. Such a fulfillment can be signitive or intuitive, in any case we call the fulfiller *evidence* (see sections 2.5.1 and 2.6). Judgment is the main theme of logic in which we investigate how from a given set of judgments other ones can be obtained. Now considering the above explained elements in regard to meaning and its fullness we can speak about the main tasks of logic, restricting us to only propositional meaning which is the ideal form of judgment.

Husserl distinguishes three levels of formal logic in his *Formal and Transcendental Logic* (Husserl, 1969, pp. 48–55):

1. Pure morphology of judgment

2. Logic of consequence

3. Logic of truth

The constitution of these three levels are interrelated, and the distinction matter if we are concerned only with formal relations. Therefore beside the three disciplines of formal logic we have a branch of study which should be called transcendental logic which investigated the roots of the mentioned themes in the transcendental subjectivity.[52] The rationale behind the above distinction can be explained, on the basis of the hitherto discussed outline of the phenomenological theory of meaning, as following. There are meanings which are obtained through empty-modification or through ideation, other meanings can be constructed following the rules of pure grammar, certain intentions are constituted whose ideal essences are these meanings, and finally these intentions can be fulfilled or not so that the meanings turn out to be full or remain empty. Now if we restrict ourselves to a specific category of meaning, namely to proposition, we can speak of three tasks: study of forms of proposition, study of derivation of a proposition out of others

---

[52] A very clear explanation of these distinctions has been given by Suzanne Bachelard in (Bachelard, 1968, pp. 11–24).

(provided that here we deal not with any proof) and the study of how the fullness of a given propositional meaning(s) grounds or grants the fullness of other given propositional meanings.

The distinction between second level, which concerns the peculiarities of derivation so that it concerns the meaning of logical connectives, and the third level, which concerns the deduction or any kind of proof as a whole in which the *truth* of the premises and the consequences are at work, is a very important one. Such a distinction remains unnoticed for the truth-functional approaches toward logic for they define the meaning of logical connectives on the basis of truth and falsity so they can not recognize the peculiarities of derivation or consequence-relation as such. Perhaps the best manifestation of this phenomenological distinction can be seen in the dialogical logic and its distinction between *play level* and *strategy level*. I will describe this in more details in the next chapter (section 3.3.2).

2.7.4 Normative standpoint toward logic

As we saw, Husserl has a broad conception of logic. Logic can not be restricted to merely operational relations among propositions. Also it is not just some supposedly true judgments beside some derivation rules. We should take into account all aspects of judgment forming and also investigate the roots of what considered to be the basic truths in regard to logical reasoning. Any truth-functional explanation of logic suffers from basic shortcomings. And also axiomatic method should be employed in a more careful manner than just taking some propositions as axioms and go through deduction. It means, the self-evidence of the alleged axioms should be examined by means of investigation on the transcendental subjectivity. All these, lead us to say that phenomenology takes a normative standpoint toward the logical systems.

I assume accordingly that no one will think it enough

2.7. HUSSERL'S IDEA OF PURE LOGIC

> to develop pure logic merely in the manner of our mathematical disciplines, as a growing system of propositions having a naïvely factual validity, without also striving to be philosophically clear in regard to these same propositions, without, that is, gaining insight into the essence of the modes of cognition which come into play in their utterance ...[53] (Husserl, 2001b, p. 248)

However it seems that during the last century, logic has been more and more taken to be a branch of mathematics as if only the mathematical notions of consistence or the set-theoretical notions of model are constitutive and central for logic. We may quote here an observation by A. J Ayer about a phenomenon in regard to the developments of logic in the 20th century:

> One of the effects is being not so much to subdue mathematics to logic, which is what Frege and Russell wanted, but to subdue logic to mathematics; and in recent years mathematical logic has become more and more mathematical and has had less and less to do with philosophy in general.[54]

However we should notice that Husserl has broad conceptions of both mathematics and logic and neither of them can be subdued to the other. However if we want to evaluate the standpoint of other schools, e.g. formalism, intuitionism, logicism etc., and do this by means of the phenomenological criteria, we should be careful about

---

[53]Ich setze also voraus, daß man sich nicht damit begnügen will, die reine Logik in der bloßen Art unserer mathematischen Disziplinen als ein in naivsachlicher Geltung erwachsendes Sätzesystem auszubilden, sondern daß man in eins damit philosophische Klarheit in betreff dieser Sätze anstrebt, d.i. Einsicht in das Wesen der bei dem Vollzug und den ideal-möglichen Anwendungen solcher Sätze ins Spiel tretenden Erkenntnisweisen und der mit diesen sich wesensmäßig konstituierenden Sinngebungen und objektiven Geltungen. (Husserl, 1913, p. 5)

[54]In the interview with Bryan Magee, which can be found here: https://www.youtube.com/watch?v=Sw1tzsMKdYQ

the terminological issues. For example they don't use the term "axiom" in a same way so that what seems to be an agreement in a first glance may happen to be a significant incompatibility. More importantly they use the word "logic" itself in different manners. As we saw above, Husserl in his preferred approach detaches logic from symbolism and from customary language. So if one accepts as the definition of logic that it is about linguistic structures, one may say, and does so in a phenomenologically acceptable way, that mathematics is independent of "logic" or rather "logic" depends upon mathematics. For this reason not only is there no basic discord between Brouwer's attitude and that of Husserl, but also they are in a substantial concurrence. Brouwer, while arguing for the independence of mathematics from logic, indeed speaks about the priority of mathematical constructions over *words*:

> The words of [the] mathematical demonstration merely accompany a mathematical *construction* that is effected without words. (Brouwer, 1975, p. 73)

And accordingly he says that words only function for communication or memory not for the construction or intellectual activity itself—the standpoint that Husserl would of course agree.

> People try by means of sounds and symbols to originate in other people copies of mathematical constructions and reasonings which they have made themselves; by the same means they try to aid their own memory. In this way the *mathematical language* comes into being, and as its special case the *language of logical reasoning*. (Brouwer, 1975, p. 73)

The accordance between intuitionism and transcendental phenomenology has been investigated and demonstrated by van Atten (van Atten, 2010). Beside the fundamental philosophical affinities,

2.7. HUSSERL'S IDEA OF PURE LOGIC

we know that intuitionistic logic rejects the truth-functional approach and this is in accordance with phenomenology's rejection of the extensional approach toward logic. Extensional approach speaks about the objectivities, and about truth and falsity as well, as if they are predetermined and references of propositions, or the value of propositional functions. Such an approach can be seen in both the algebraic method, which has been criticized by Husserl even in the early period of his career (Husserl, 1994a) (Husserl, 1994b), and the axiomatic method of Frege and Russell. They do not pay attention to the process of experiencing truth, and thus pay no attention to the intensional feature of logical elements. For Husserl this is the act of striving toward truth which is essential, and this is what phenomenology and intuitionism both strongly agreed upon. Any logic which defines the logical elements only on the basis of some extensional relations, or the function between truth and falsity as extensionally given values, could not reveal the essence of logical issues.

We can see that phenomenology would not naively take for granted the validity classical logic. The fact that Husserl has not set for himself the task of providing a phenomenologically acceptable logical system does not mean that he would accept the science of logic as it is given.

Certain scholars, based on some textual evidence from Husserl, state that in his view philosophy is not in the position to revise a science—including logic. This stance called by Lohmar (Lohmar, 2004) "conservative". But we are in agreement with Lohmar himself who following a suggestion from van Atten (van Atten, 2002) adopts the term "revisionism" to describe Husserl's attitude concerning the relation of phenomenology with sciences—logic and mathematics in particular.

The need for revision of logic can be seen in the same line that Brouwer speaks of the unreliability of some logical laws. In his 1908 paper *The unreliability of the logical principles*, Brouwer asks:

> Can one, in the case of purely mathematical constructions and transformations, temporarily neglect the presentation of the mathematical system that has been erected, and move in the accompanying linguistic building, guided by the principles of the *syllogism*, of *contradiction*, and of *tertium exclusum*, always confident that, by momentary evocation of the presentation of the mathematical constructions suggested by this reasoning, each part of the discourse could be justified? (van Atten and Sundholm, 2016, p. 25)

As it can be seen his doubt is about the validity of principles of *linguistic buildings* in respect to pure reasoning. It implies that a syntactic consistence by itself is not sufficient and in order for a principle to be genuinely valid something more is needed. Brouwer in that paper shows that the law of excluded middle is not generally valid, though, as van Atten and Sundholm discuss in their introduction to the translation of the mentioned paper, he observes that this law is consistent (van Atten and Sundholm, 2016, p. 9). It means that a mere consistence is not enough for a logical system; and a revision, in accordance with the more basic features of intellectual activity, is required.[55]

Therefore, the course of phenomenological meaning-explanation of the logical constants involves evaluations of the given logical systems or introduction of some new rules. As it is to be explained in the next chapter, the dialogical semantics provides an appropriate, formal framework to carry out such phenomenological meaning-explanations of the logical constants and evaluations of the logical rules.

---

[55]In particular, Brouwer's refutation of LEM, is granted by the phenomenological method. I will argue for this in chapter 4.

Chapter 3

Phenomenology and Dialogical Semantics

3.1 Links between the Phenomenology and Dialogical Approach

In section 1.6, certain phenomenological themes in respect with the alterity were briefly discussed. There we saw that how alterity is connected to the horizontality of intention and to the conception of the other as an ego-modification. A dialogue, whether inner or outer, can be analyzed by means of the mentioned phenomenological concepts. There are of course huge differences between the outer dialogue and the inner dialogue; however as far as the logical issues are concerned we may restrict ourselves to the analyses of the inner dialogue. The reason is as follows. The main difference between the outer dialogue and the inner dialogue, from which the other differences follow, is that in the inner dialogue I have my actual standpoint, to which all my positional acts, including the unmodified judgments, are related, besides some hypothetical or fantastical standpoints, to which some quasi-positional acts like modified judgments are related. So, in a concrete situation I always have a particular standpoint at the time of reasoning. On the other hand, in an outer dialogue, when observed from the outside, there are two "actual" standpoints, so that we have two genuine perspectives to the objectivity in question and we may have positional acts and unmodified judgments from the both sides of the dialogue. Nevertheless, when we are dealing with the ideal objectivity, like that of logic and mathematics, the concrete situation is irrelevant to any judgments about the objects, so that the difference between the actual standpoints does not matter. Then here, as opposed to a dialogue on a material objectivity, the difference between the inner dialogue and the outer dialogue has nothing to do with the process of reasoning as such.

In the present work, I wish to put some ideas of the dialogical logic, introduced by Paul Lorenzen, in the line of the phenomenological method in order to reach a comprehensive framework for logic and to explain the meaning of the logical entities as well. In this respect, as I have argued for, the notion of the inner dialogue would be central. I do not aim to investigate the connections between the phenomenological method and the dialogical approach in general. However, here I will review a book by a well-known phenomenologist who aims to reconstruct the phenomenological method by means of the dialogical approach. Through the study of the that work and discussing some ideas of it I will be able to make clear some peculiarities of the dialogical approach as recalled in the present project.

3.1.1 Stephen Strasser's book: *The Idea of Dialogal Phenomenology*

"The Idea of Dialogal Phenomenology" (Strasser, 1969) is the effort of its author to revise the fundamental principles of phenomenology in order to make it able *to solve the problem of intersubjectivity*. Strasser takes it as the starting point that Husserlian phenomenology

3.1. STRASSER'S BOOK

fails "to account for the social dimension of the human existence". This seems to be a basic assumption of his work that intersubjectivity appears for Husserlian philosophy with its transcendental subjectivism as a problem. In this note I will provide a critical review of this book.

In the preface, the author implicitly makes a distinction among problems for a given philosophical method: between those that are supposed to be dealt with as the method advances and those that are challenging the very possibility of beginning of that method. That is to say, it is not problematic that for a philosophical method—and phenomenology is not an exception—there can be recognized or introduced problems for which there are no already available explanations within that method. Nevertheless, another type of problems may exist that, if they remain unaddressed, would put into question the legitimacy of that method itself.

The author states that for a long time he has been thinking that "Husserl's failure to account for the social dimension of the human existence is a problem of the first kind" (Strasser, 1969, p. xii), but later he recognized that this is indeed a fundamental problem so that a revision of basic theses is necessary; a revision that takes the intersubjectivity, the dialogue among humans, as a paradigm from the start. So, the book is to be read as a response to this need.

Therefore, in the book, the idea of dialogue does not appear as a matter under study but as a key concept in regard to which the main themes of phenomenology such as consciousness, intentionality, horizon, freedom etc. are to be reconstructed.

The book contains six chapters. In chapter one, the author introduces and explains four characteristics of Husserlian phenomenology. These characteristic elements, according to the author, are intuition, reduction, intentionality and constitution. Explaining them, the author wishes to convince the reader that there are inconsistencies in the basis of phenomenology, which therefore is not reliable in this given manner and calls for revision. In the second chapter, the notions of "world" and "horizon of experience" are analyzed. In the third chapter the main idea of what is called "dialogal phenomenology" is explained. The fourth chapter is dedicated to the psychology of consciousness. Here, the author tries to demonstrate the role of the I-You relation in the forming of human consciousness. A similar analysis is carried out in the fifth chapter about the "freedom". The sixth and last chapter contains analyses about what the author calls "faith".

The book includes interesting studies about the forming and the effects of certain elements of consciousness in human life, especially during the course of growth. Thus, if the reader has been convinced with the arguments in the preface and the first chapter about the necessity of revision in phenomenology, he would find the presented studies quite relevant and fruitful.

In the following I will represent a short report of the content of the book, which is followed by a brief discussion.

* * *

1- The first chapter aims to clarify what, in the eyes of the author, are the specifications of the phenomenological analysis. These are four in this order:

1. Intuition. Phenomenological analysis is a method based on intuition, not speculation. That is to say, the phenomena are supposed to be "seen" as they are.

2. Reduction. An important doctrine of phenomenology is *epoche* by which the ego strives to know the world from the standpoint of a disinterested observer. Therefore, all *real* connections between the ego and the phenomenon under question and thus all judgments about the existence of the phenomenon are suspended. The phenomenon is to be reduced to its essence— in contrast to its existence or existential values.

3. Intentionality. According to phenomenology, intentionality is

3.1. STRASSER'S BOOK

the principal property of consciousness, so that any analysis pertaining to consciousness is essentially intentional analysis. After executing reduction of an objectivity, what remains is the intention toward it. The author rightly points out the relation between intentionality and the two previous notions, and also that with the notion of constitution, which is the fourth element.

4. Constitution. A large part of phenomenological studies deals with the constitution of objectivities. The author prefers to call this notion by the term "genesis" which poses no problem, provided we have a phenomenological account of this term. However we should notice that the phenomenology of consciousness stands, in some sense, in contrast to genetic analysis of consciousness, if by "genesis" one intends a specifically natural-scientific notion.

While explaining these theses, the author points out what he takes to be deficiencies of the transcendental phenomenology. The main shortcoming is the contradiction he sees between the emphasis on intuition and the employment of the reduction. As Strasser puts it, it is contradictory that on the one hand we acknowledge intuition as the ultimate source of knowledge of objectivities and on the other hand by reduction we avoid accepting the phenomenon as it appears.

> Nevertheless, the reductive ethos of transcendental phenomenology demands that we not surrender ourselves to the spectacle of things. For this would mean giving in to the "natural attitude." The naiveté of lingering, in astonishment and admiration, with things, living beings and persons—in short, with the world—needs to be overcome by the phenomenological reduction. Through an act of asceticism we discover an entrance to a new realm, the realm of psychical consciousness with its meaningful

noeses and *noemata*. This psychism could also be a region in which we could more deeply immerse ourselves through our acts of internal contemplation. Yet, we must once again tear ourselves away from it, for the being of that "inner world" is, we are told, just as much based on the achievements of transcendental subjectivity as is the being of the "external world."...

Once again, then, the phenomenologist who expects finally to satisfy his desire for insight through intuition meets with disappointment. He is not unlike Sancho Panza, the man who was called to be king in Cervantes' famous story: Sancho was offered the most appetizing foods, but as soon as he wished to eat of them, the royal physician came foreward, declaring that the dishes were unhealthy and ordering that they be taken away. (Strasser, 1969, p. 9)

Accordingly, Strasser finds the idea of "disinterested observer" quite unsatisfactory. After some discussion he concludes that *the two tendencies—confidence in what is given and the Cartesian search for apodictically certain knowledge—did not reach a perfect balance in Husserl's philosophy.*

Furthermore, the author lists what he calls inner tensions of phenomenology. These are:

1. "Phenomenology presents itself as a philosophy of intuition. But, on the other hand, it is a philosophy that relativizes and brackets what is given through intuition." (Strasser, 1969, p. 18)

2. In the process of reduction, phenomenology leaves the concrete facts in behalf of abstract postulates. So, this is inconsistent with its claim in being faithful to the experiences as such.

3.1. STRASSER'S BOOK

3. Through intentionality one is concerned with what is not itself consciousness. This is in conflict with the doctrine of disinterested observer.

4. Phenomenology desires to describe the constitution of the world by studying the consciousness. But it would be infeasible if we emphasize on the *a priori* character of the consciousness.

Strasser concludes the first chapter by declaring what he sees as the cause of these claimed inner tensions.

> We shall have to renounce two deeply ingrained prejudices. We shall have to realize that phenomenology cannot begin with an ego-logy and secondly that reflection cannot be its principal method. (Strasser, 1969, p. 19)

Accordingly, he declares the task of dialogical phenomenology[1] as such:

> The task of a dialogal phenomenology will be to describe how a world arises for us in the dialogue between me and the other. It will also examine the turning points of this dialogue and the corresponding changes of worlds. (Strasser, 1969, p. 22)

2- In the second chapter, Strasser attends to a difficulty in defining "I" as it appears in the I-You relation. For him the other side of defining "I" is defining a "world" in contrast to which and within which "I" exists. He cites some notable phenomenologists who explicitly define "I" by referring to a notion of "world".

---

[1]Strasser uses the term "'dialogal phenomenology". However in this writing I prefer to use the more common term "dialogical" in order to speak of a philosophy which aims to take dialogue as the paradigm.

Merleau-Ponty declares that "I am only a center of worldly situations"[2]; and Sartre considers man's relation to the world as the concrete starting point of his philosophizing.[3] It is possible, then, that the phenomenologist's answer to the question, Who am I?, includes a reference to the world.

Then the author mentions and examines most plausible meanings of "world", as "the totality of beings for us", "structured reality" or "horizon of all experiencing intentions" (Strasser, 1969, pp. 24–26).

On the basis of the possible meanings of the concept "world", Strasser points out the difficulties of speaking about a world-less life as it is claimed in the transcendental philosophy. These are indeed very serious difficulties[4], whether or not the reader agrees with the author's conclusion about the impossibility of reduction. However, the author declares his standpoint and states why for him dialogical philosophy is preferable to the transcendental phenomenology. (Strasser, 1969, p. 45)

3- In the next chapter, the author tries to clarify his account of dialogical philosophy. His effort is dedicated to reintroducing some important philosophical themes in the light of dialogue. In this chapter we encounter interesting topics like: *A Pluralistic View of the World Problem, How is a "You" for "Me"?* and *Dialogal Constitution*. Perhaps the central argument lies in the author's explanations of the intention toward others and how a "You" functions in the intentionality of the ego. He emphasizes the reciprocity of relations in the ego's living experiences.

---

[2] *Phénoménologie de la perception*, p.520.

[3] *L'être et le néant*, p.38.

[4] However, all difficulties about the possibility of epoche and disinterestedness in the world, I think, are difficulties for who examines the transcendental philosophy from within the natural standpoint, for in the natural life there is no motivation to suspend it, thus such an examiner hardly distinguish between suspension of the world, a transcendental move, and denying the world, which is absurd.

3.1. STRASSER'S BOOK

> If we understand by a subject a being that is exclusively active, and by an object a being that is purely passive, then no concrete relations between the two are conceivable. For in what would those relations consist? Every concrete activity—including every activity of knowing—demands, according to our experience, something that offers resistance to it. But where would that resistance come from? Every striving—including that for knowledge—overcomes obstacles to reach its purpose; thus, again, the possibility of resistance is presupposed. (Strasser, 1969, p. 55)

That is to say, in my experiences, I don't find myself as absolutely active and the objectivity in front of me as absolutely passive. Rather there is a reciprocity being realized in an I-You relation. According to Strasser's argument, the "you" is the being which is originally present to me, which is not the object of my activity but first that of my belief. Indeed Strasser goes further:

> my affirmation of the "you" must transcend all doubt for me; it must be characterized as the "primordial faith" (Urdoxa) upon which all my further *cogito's* rest. (Strasser, 1969, p. 61)

Therefore, every experience, Strasser claims, is based on an I-You relation, for

> It is precisely through the mediation of a "you" that I know at all that there are things worth touching, tasting, looking at, listening to ...

The I-You relation, nonetheless, is not a fully clear idea; thus Strasser tries to clarify his stance and thereby explain his account for the important notions of "cogito" and "constitution". (Strasser, 1969, pp. 62–68)

4- Chapter four contains Strasser's studies about the originating of consciousness in virtue of the I-You relation. Most great philosophers have contributed to the discussion of this subject and the differences between their views are by no means peripheral. In order to declare his own approach to the problem, Strasser mentions some viewpoints asserted in the context of phenomenology. Indeed in this chapter he refers to numerous works of contemporary philosophers and also psychologists. His main attitude, one may recognize, is that of developmental psychology. In this respect he widely uses the contributions of most notable psychologists of the time.[5]

However, this chapter contains a lot of original discussions in the issue. Nonetheless, as well as these discussions are original, they are questionable from the phenomenological point of view. I will come back to this in the last part of this review.

5- After the problem of awareness, the author turns to another important topic in the human life, namely *freedom*. He studies the relation between freedom and dialogue and through this study he wants to shed a new light on the concept of freedom itself.

Strasser sees the idea of absolute freedom untenable (Strasser, 1969, p. 105). He simply thinks that this is in contradiction with our experiences as human beings; and accordingly he finds the attribution of infinity to freedom so strange that he, instead of analyzing this idea itself, prefers to explain the historical rout which, in his eyes, has led to the birth of this idea.

The main point in his argument is that determinism is not the only alternative for absolute freedom; we can speak of finite freedom though it seems odd at first glance. Taking the dialogue as paradigm, Strasser argues, we can conceive how finite freedom is possible, where the subjects have to adjust themselves to their partners and to the object of the dialogue, "in a way that an understanding and

---

[5] Among the scientists to whose work Strasser refers we can name Piaget (Piaget, 1947), Remplein (Remplein, 1966), Spitz (Spitz, 1992), Store and Church (Stone and Church, 1957).

3.1. STRASSER'S BOOK

agreement with the interlocutor is not *a priori* excluded." (Strasser, 1969, p. 102)

Thus the author tries to defend the idea of finite freedom in the sense that the freedom of every human being is essentially restricted by the freedom of others.

6- The last chapter is dedicated to the ontological dimension of dialogue and the role of beliefs towards existence in the human life. The author deals with the idea of faith as it functions in relation with a "You" and thus functions in various stages of life. He explains this latter under the title *Dialectic of My Encounter* (Strasser, 1969, p. 131). He used here the term "dialectic" because he thinks that the process of encounter in the course of life passes through the phases of "belief", "unbelief" and "believing recognition"; but his explanations here is incomplete, it is a short, quite debatable narrative for what that was supposed to be "a metaphysical reflection upon the course of human life".

The book concludes with a remark that to believe in the existence of the "You" is the fundamental attitude of man (Strasser, 1969, p. 132).

* * *

In order to analyze the content of the book, I will discuss the following three questions:

1. Regardless whether the contents of the presented studies are correct or incorrect, is such a project able to satisfy its original aim? In other words, can these analyses be considered as an alternative to what the author calls monological analyses?[6]

2. Are the methods of investigation employed in this project truly phenomenological? To what extent the methods used in the

---

[6]In fact, Strasser uses the term "monologal" to indicate that kind of analyses which begins with an ego-logy and its principal method is reflection. (Strasser, 1969, p. 19)

book are phenomenologically acceptable?

3. Isn't there any other main feature, beside those studied in this work, belonging to "dialogue" that must be necessarily dealt with when we want to establish a philosophy on the basis of dialogue?

1- I begin with the first question. We can be in complete agreement with the author in saying that the problem of intersubjectivity is a challenging one for phenomenology who widely uses an egological method.

Phenomenology is supposed to commit itself to the phenomena in lived experiences; and the experience of living with others is one of the most important. Thus it is a quite reasonable expectation that phenomenology should give an explanation of the possibility and the status of intersubjective life compatible with its foundations. Strasser might be right in thinking that phenomenology had not brought forward a convincing explanation.

However, should this be considered as a fundamental problem, as the author claims, or as a problem-in-path? Strasser does not give his reasons for giving up his initial stance and considering this problem as a fundamental one rather than as one to be solved as the method progresses. He rather connotes that a philosophy which does not methodologically begin with intersubjectivity must be problematic from the outset. So, a phenomenologist may find the content of the book not so relevant to a genuine problem within phenomenology; rather a phenomenologist may find it interesting but not in regard to its primary claim. He/she may still think that the problem of intersubjectivity is not of that kind that brings about a necessity of revision.

There is also a third possibility which I would like to support. That is, the problem of intersubjectivity is not a problem-in-path, but yet in order to give a phenomenologically acceptable treatment of the problem there is no need for a basic revision of this method.

3.1. STRASSER'S BOOK

Rather we should add certain principles to it; or to put it in other words, we should adopt certain principles, namely monadological principles, before entering the path of phenomenology. So that the structure of transcendental phenomenology without being basically changed is reestablished in the more thorough philosophy of transcendental monado-phenomenology.

Husserl himself has repeatedly speak of the affinity of the transcendental phenomenology with monadology. Thus there is no problem in principle to incorporate phenomenology, at least as its founder conceived it, with monadology. Husserl has even claimed that "phenomenology leads to the monadology anticipated by Leibniz" (Husserl, 1959, p. 190). But I think this is not the case if it is meant by it that the systematic development of phenomenology results in monadology. That claim is, however, quite true if it means that advancing in the method of phenomenology, the philosopher will reach a point to understand more than any other the depth and necessity of a monadological metaphysics.

One may acknowledge, alongside with Strasser, that by progressing in the method of phenomenology we do not reach a comprehensive explanation of intersubjectivity, that is, we do not reach transcendental monadology. Given this be true, this point does not prevent a self-consistent development of phenomenology beginning from the monadological bases to arrive comprehensive analyses of intersubjective experiences and experiences of intersubjectivity.

So, I think there is no reason to seek for inconsistencies in the foundations of phenomenology, as Strasser does, in order to justify its disability, in its current state, in appropriately meeting the problem of intersubjectivity.

2- While investigating a sequence of attractive topics such as awareness, freedom, faith etc. there is an issue which appears unsatisfactory to a phenomenologist reader of the book. That is to use the methods close to psychologism and naturalism in most parts of the studies. For instance, most of Strasser's arguments are based on developmental psychology. But such arguments, though themselves interesting in some respects, should not be referred as fundamental phenomenological studies.

Although Husserl pays much attention to psychology and takes phenomenology's contribution to this science for serious and he even repeatedly speaks of the parallelism between phenomenological psychology and transcendental phenomenology (Husserl, 1960, §57) (Husserl, 1969, p. 15) (Husserl, 1988), this by no means entails that the former (in any possible form) can substitute the latter or even plays an elementary role in the investigations peculiar to it, namely investigations on constitution, objectifying, awareness, freedom and so on.

Husserl, from the time of *Logical Investigations* on, in the whole his philosophical project bewares of any kind of natural-scientism and specifically of psychologism; or, as he himself says in a lecture from 1931, his answer to "the question of whether there can be any legitimacy to a philosophy whose grounding rests on the essence of human beings in any form whatever" is certainly negative; for "it is immediately clear that any doctrine at all of human being, whether empirical or apriori, presupposes the existing world or a world that could be in being." (Husserl, 1988) while this presupposition is the falling back to the naivety of the natural attitude.

However, the aforementioned feature of Strasser's approach is not inconsistent with his own position, for he has shown from the start that he does not accept the distinction between natural and transcendental attitude. Then it is not strange to see that naturalism frequently appears in his work. But we can ask then in what sense this project is phenomenological. The answer that the author gives is not satisfactory. He says:

> It is true that dialogal phenomenology in many respects differ from "classical" phenomenology. Its principal theme, however, is preserved almost intact. It is the

3.1. STRASSER'S BOOK

same that Eugen Fink formulated in 1933 with Husserl's approval.[7] The essential question of phenomenology is and remains that of the origin of the world. (Strasser, 1969, p. 22)

It is true that phenomenology deals with the question of the origin of the world, but to deal with a question is too ambiguous to define a philosophical system as such. A philosophy is first of all determined by its method, and we know phenomenology by *epoche*, by *eidetic variation* and by focusing on transcendental subjectivity, but Strasser's studies are widely based on factual issues about the development of the natural mind.

Just to clarify my criticism, I want to mention a discussion from the book and I choose the topic of *naming*. Now let us see Strasser's analysis of the subject matter. I here bring the summary of his argument.

By giving names to things, the child definitively transcends the emotional relationship of the first phase and the purely practical relationship of the second phase. Let us illustrate this point by means of an example.

The sound "Mama" was for the little child first a cry of joy. Later "Mama" became the expression of a practical desire. It meant, for example, "I should like to be picked up." But finally "Mama" becomes the name of a person, albeit a person of particular importance for the child. Now the word no longer has a purely expressive, or purely practical meaning for the child; it now has an intentional function. This means that the child's consciousness through and beyond the sound of "Mama" is directed to the person of his mother. (Strasser, 1969, p. 93)

---

[7]"Die phänomenologische Philosophie Edmund Husserls in der gegenwärtigen Kritik". von Dr. Eugen Fink. Mit einen Vorwort von Edmund Husserl. *Kantstudien*. Berlin. 38 (1933): 319-20.

In the part of the book from which the above phrase is taken, Strasser introduces a thesis about originating of the intention of naming which is in principle one of the more important intentions constituting expression. Such a thesis should surely be studied and examined but the question is that to what extent this thesis has prepared itself to meet phenomenological examinations. Is the presented thesis really free from those presuppositions that phenomenology is supposed to avoid?

The problem is that those three stages claimed by Strasser, i.e. emotional phase, practical phase and objectification by means of names, can not be intuitively evaluated and are of speculative nature on the basis of scientific theories. Such a wide use of psychology is not phenomenologically acceptable. To present a natural thesis about the origination of awareness is in conflict with phenomenology not only because it is explicitly in contrast with the transcendental attitude but also because it would necessarily be based on natural sciences and theoretical claims and thus suffers from unclarified prejudices.

So, for a phenomenologist it is very bizarre when, for example, he is confronted with this assertion:

> About three years of development finally make the child be what he thought to be according to classical phenomenology: he is now awareness of a world. (Strasser, 1969, p. 94)

There is more to be said about the flaws of the natural attitude used in the book by Strasser, but here I only want to show that he, though deliberately, works in this attitude; and this makes the project hardly appropriate for who have already been convinced by the arguments that the transcendental phenomenology offers about of the role and limits of the natural attitude.

3- The third remark is about the issues that were expected that the author deals with since he speaks about dialogue. Although we

3.1. STRASSER'S BOOK

see in some places insightful analyses of dialogue[8], in most places it is spoken of dialogue as it is indistinguishable from the general case of interaction. The aspect of *logos* in the act of dialogue has been mainly neglected. There is no discussion about language except that short part about naming which I mentioned before. To refer to the idea of dialogue without a sufficient discussion about language could not provide a cogent ground to start.

I think that the argument about the possibility or impossibility of inner dialogue is also a decisive issue. For if I always conceive the other as other or as You in an I-You relation a considerable aspects of dialogue will be lost. One is the possibility of empathy, namely dialogue with You as a possible I, as I-from-that-viewpoint. How this You as an Other-I would be explained in the Strasser's system? For he puts the experiences of man from the outset in the model of I-You, so that always You appears as You regardless of having any essential character in common with me. Indeed, Strasser frequently states that "you" is always older than me (Strasser, 1969, pp. 52–61), it is a "giver" not a "given" (Strasser, 1969, p. 62). Thus, according to him, "you" poses itself always as "you". But, we should notice that in our primordial experience of the I-you relation, it is a substantial feature that I confront the "you" also as an "I", not as that I which I actually am but as an "I" having those experiences, those visions, those perceptions, that I could possibly

---

[8]For instance, while studying the laws of the true dialogue, he introduces the following conditions:

1. The one who participates in dialogue "must adjust himself to his partner in the dialogue by listening and answering, by affirming and denying, by doubting and agreeing."

2. "He must take a stand in relation to the "something" that is object of the dialogue."

3. "The subject who is engaged in a dialogue must relate himself to object under discussion in such a way that an understanding and agreement with his interlocutor is not *a priori* excluded." (Strasser, 1969, p. 103)

have but of their contents I am not actually aware.

In order to understand the idea of Other I which is of a prominent importance in dialogue, we should attend to the idea of inner otherness. If I and Other appear from the start as separated and from distance whose interactions, each from its own place, form the life, there would remain no possibility to assume Other as a possible I. The proper solution indeed is that the concept of otherness has been already originated in the living experience of I, so that it be a ground for the idea of Other I.[9]

I agree that there is *a* relation of I-You which is by no means derived from the experiences of a self-sufficient ego, the relation in which I don't, and cannot, confront the "you" as a possible I but as the absolute "You". This relation, nevertheless is not the general form of the intersubjective relation but in fact it is unique to the I-God confrontation.

The relation in which "you" is substantially *giver*, or is always *older than me*, or my affirmation of it "must be characterized as the "primordial faith" (*Urdoxa*), upon which all my further *cogitos* rest" (Strasser, 1969, p. 61), is not the kind of relation that is realized in the concrete social life. This is a unique relation which is meaningful only in I-God relation and has been abstracted from it.

The arguments that the author carries out under the title of "faith" are merely related to "belief" and has nothing to do with faith in its existential meaning; namely an existential orientation of the self which is not based on any objective foundation that is external to the self. But solely in this latter sense phenomenology may acknowledge "faithfulness as the fundamental attitude of man"[10]; otherwise and if by faith one means belief, as Strasser does, phenomenology can never admit it as a fundamental attitude[11].

---

[9]For more discussion on the inner alterity and its importance see sec. 1.6.

[10]The expression between quotation marks is Strasser's (Strasser, 1969, p. 132)

[11]See also our discussion in sec. 2.6.2

3.2 A Short Introduction to Dialogical Semantics

> *All deliberative meditation, or thinking proper, takes the form of a dialogue. The person divides himself into two parties which endeavor to persuade each other. (C. S. Peirce)*[12]

The dialogical approach introduced by Lorenzen and Lorenz is a successful elaboration of this idea. The dialogical approach provides us with a semantics which is not model theoretic nor proof theoretic.[13] Dialogical method distinguishes between two kinds of logical rules, 1- the rules governing local moves, namely the rules determining that how each form of complex formulas can be attacked and defended, which are called particle rules, and 2- the rules governing the whole dialogue which determines the rights and the obligations of each party and that how a dialogue proceeds and terminates and who is the winner, these are structural rules. It accordingly distinguishes between the so-called play level, which is the moves of derivation, and the strategy level, which is to order a chain of such moves in order to reach the conclusion. The point is that such distinctions makes us able to deal with meaning and truth separately. The method grants that the meaning of a logical connective is not confused and is not dependent on the truth or falsity of its terms.

The original motivation of dialogical semantics was to cover intuitionistic logic. The reason should be clear from the above explanations. However it was proved to be capable to cover a considerable range of other logical systems.[14] Moreover, using the dialogical framework one would have a new insight into the

---

[12] In (Pietarinen, 2015, p. 7).

[13] For a good explanation of this point see (Rückert, 2001).

[14] For an study of the various dialogical semantics for different logical systems see (Rahman and Keiff, 2005).

conceptual, not merely technical, differences of alternative logical systems.

I bring a short introduction of the dialogical semantics in the following.[15] Then, in the rest of this section I put forward phenomenological analysis in order to explain the philosophical importance of the main themes introduced by the dialogical framework.

In the dialogical semantics there are two parties, the proponent **P** and the opponent **O**. The proponent introduces a thesis and defends it against the attacks of the opponent. If there is a winning strategy for the proponent in respect to a statement, that statement is valid. The attacks and responses are to be performed according to two kinds of rules, particle rules and structural rules.

Particle Rules For any logical connective there is a particle rule which determines how to attack and defend a formula with a specific main connective. These rules are standard in the literature:

	Attack	Response
$A \vee B$	$?_\vee$	A, or B
		(The defender chooses)
$A \wedge B$	$?_L$, or $?_R$	A, or B
	(The attacker chooses)	(respectively)
$A \to B$	A	B
$\neg A$	A	(No possible response)
$\forall x A$	$?_{\forall x/c}$	$A[x/c]$
	(The attacker chooses c)	
$\exists x A$	$?_{\exists x}$	$A[x/c]$
		(The defender chooses c)

---

[15]The following rules are standard within dialogical studies. However in the current manner of presentation, I particularly benefited from the representations given in (Rebuschi, 2009) and (Rahman and Keiff, 2005).

3.2. A SHORT INTRODUCTION

Structural Rules Structural rules determine the structure of the interaction which form a certain argumentation. We have:

(SR-0) Starting Rule: The Proponent begins by asserting a thesis.

(SR-1) Move: The players make their moves alternately. Each move, with the exception of the starting move, is an attack or a defense.

(SR-2) Winning Rule: Player **X** wins iff it is **Y**'s turn to play and **Y** cannot perform any move.

(SR-3) No Delaying Tactics Rule: Both players can only perform moves that change the situation.[16]

(SR-4) Formal Rule: **P** cannot introduce any new atomic formula; new atomic formulas must be stated by **O** first. Atomic formulas can never be attacked.

(SR-5c) Classical Rule: In any move, each player may attack a complex formula uttered by the other player or defend him/herself against *any attack* (including those that have already been defended).

(SR-5i) Intuitionistic Rule: In any move, each player may attack a complex formula uttered by the other player or defend him/herself against *the last attack that has not yet been defended*.

---

[16] In order to observe this rules it should be determined that each party how many times may repeat a same attack (repetition of defense is redundant in any case and it is not allowed). This is called *repetition ranks*. Clerbout (Clerbout, 2014) has shown that it would be sufficient to assign the rank 1 to the opponent and 2 to the proponent, namely there would be a winning strategy for a formula if and only if there is a winning strategy for that formula while the proponent is allowed to attack twice against a same move and the opponent is allowed to do so only once. Therefore, I do not specify the ranks in the following dialogues, and one can suppose that it is 1 for **O** and 2 for **P**.

The mentioned structural rules pertain to the *formal* dialogue, namely a dialogue to examine a formal truth, i.e. a logical validity. If we want to examine a logical consequence of some given premises, the rule (SR-4) should be modified in this way that the proponent may introduce any proposition given in the premises and also may ask the opponent to assert a specific premise—that the proponent chooses from the set of premises which are agreed upon before entering into dialogue. Such a kind of dialogue are called *material* dialogue. In the rest of this project we will deal with formal dialogues, but it is good to have in mind that what we are discussing can be applied also to material dialogues.

Perhaps it would be good to give a simple example to show how this semantics works.[17] Let us examine the formula $p \to (q \to p)$.

> Remark on notation: The moves are brought in the order of utterance. The parenthesized information indicates the number of the move, whether it is an attack (ⓐ) or a response (ⓡ) and to which move it is an attack or a response. The long dash indicates that there is no further move; and the player has been lost.

	O		P	
			$p \to (q \to p)$	(0)
(1ⓐ0)	p		$q \to p$	(2ⓡ1)
(3ⓐ2)	q		p	(4ⓡ3)
	—			

Here in the move 4, **P** is able to respond the attack, because **O** has asserted p before; and since there is no other move possible for **O**, **P** wins and the formula is demonstrated to be valid.

---

[17]In the following dialogues I just sketch out the core play which can be used to build a winning strategy. For the sake of simplicity I do not discuss all possible plays to show that in any case there is a winning strategy for **P** (or for **O** if I show the thesis is not valid); it is not difficult to show and I omit it in order to focus on the core of the argument.

3.3 Phenomenology of Logic and the Dialogical Approach

3.3.1 The transcendental status of dialogue: the idea of modification

In this investigation I take the phenomenological idea of *modification* as a guideline in order to clarify the meaning of logical connectives and the status of logical rules. First of all let us recall the distinction between proposition and judgment. The plain form of judgment is truth-judgment which bears certitude; but other types come to stage primarily as modifications, so we have negation, supposition, possibility and so on. In any case proposition is the ideal form of the judgment; and a proposition of the mentioned types can be constituted either by an empty-modification of the corresponding judgment (sec. 2.2.3) or by means of the categrial synthesis out of the simpler meanings (sec. 2.3). Since in argumentation we primarily deal with judgments a great task of logical investigations is to clarify the judgment modifications, their relation to each other and their role in the course of argumentation. Here I just want to show the relation of this philosophical background with the dialogical approach.

Husserl recalls the notion of modification, most importantly, while discussing these topics: judgment and its modifications, the awareness of the other as a modification of "I" (Husserl, 1960, p. 115), the account of the past and the future as modifications of the present. My idea is to link between the judgment-modification and the inner ego-modification. That is, an argumentation, in which different types of judgment-modifications are supposed to be at work, can be understood as a course of inner dialogue between different ideal types of I-modifications, or more precisely between a central "I", who is supposed only to assert genuine judgments, and an inner I-modification who is in charge to assert different judgment-modifications according to certain eidetic rules.

Therefore, the primordiality of the inner dialogue is essential to the presented idea to incorporate the dialogical semantics within the phenomenological method. Any sociological or anthropological account of dialogical activity is irrelevant to the present investigation, as it is irrelevant to any investigation on pure logic.

I will elaborate the presented interpretation in the following. But in order to have a preliminary idea I should say, here we have two parties one is an unmodified "I" who intends to assert the unmodified judgment namely to strive toward truth, this is the proponent, and the other is an *inner other* or a modified "I" who is allowed to assert modified judgments and thus to perform assertions regardless to the genuine evidence while confronting the central "I", this is the opponent. In respect to every kind of judgment-modification, or correspondingly every propositional constant, there is a pattern of interaction according to which one may challenge the other or defend. These are particle rules. There are also some general conditions each of the parties must observe if the overall intention is to demonstrate the truth. Here, in contrast to the former set of rules, we have asymmetric rules, that is, this latter set of rules concerns the role of each party in the argumentation and the feature of the argumentation as a whole. These are structural rules. Now having an idea of the main proposal we further to develop the thesis in details.

3.3.2 Play level vs strategy level as related to consequence-logic vs truth-logic

The dialogical method distinguishes between particle rules and structural rules, and in relation to this distinguishes between *play level* and *strategy level*. The validity of the argument as a whole, or equivalently the construction of a proof, is determined in the strategy level, that how the proponent proceeds in defending his

3.3. PHENOMENOLOGY OF DIALOGICAL LOGIC 203

thesis. However regardless to whether an argumentation is ultimately valid or not we can determine whether it proceeds in a properly logical way namely the logical connectives are treated as they should be; this is the play level. In the following I will try to elaborate the distinction and show how it may contribute to our understanding of logic.

Formal logic, according to Husserl, has three levels (sec. 2.7.3). The first level is the study of pure morphology of judgment or, generally, of various grammatical forms of expression which are used in reasoning. The second level is what Husserl calls consequence-logic. The third level is truth logic. The distinction between the latter two levels is very important while trying to clarify the nature of logical argumentations.

Let's think about the logical connectives and the role they play according to their own meaning without considering the relevance of truth to their terms. Namely, consider the connectives only as a kind of operation, or as Husserl calls it, mere analytical includedness and excludedness. For instance, the following can be read in two ways: $p \wedge q$ yields p. In a reading it serves as a truth-preserving relation and itself corresponds to a small proof. But also we can read it without considering that the terms are subjects to truth, but as an empty derivation or an operation which works on propositions so that the relation merely shows that how conjunction works. As far as we are working within deduction the distinction is not that much relevant. However, once we go to dealing with semantics and explanation of logical relations the difference between two readings turns out to be important.[18] In order to have a truth-logic, beside

---

[18]Nevertheless, not only within semantics but also in the axiomatic method when we deal with constitutional problems the distinction is very relevant. For instance if we take the mentioned formula, $(p \wedge q) \to p$, as an axiom, we can read it as a fundamental truth, as Husserl would do, or as a mere schema, as Hilbert does. Then, although both Husserl and Hilbert defend the axiomatic method, their approach are radically different from the philosophical point of view, for Hilbert the axioms are but schemata, the reading which accords his

the derivational feature of logical connectives we need also some "determinant" rules, for instance the repetition of premises is allowed or not or absurdity should be avoided or any proposition is true or false etc. Such rules do not reflect the characteristic of the logical connectives but the characteristic of the general ontology, or correlatively that of the thinking subject, to which and by which logic is applied. The principle of identity, for example, or that of non-absurdity are metaphysical principles at the first place (one can list here the alleged bivalence and the like), without which we are not able to speak about *proof* but a mere derivation. However it is important to notice that these two levels are not completely separated so that each of them can be developed independently. Rather this is in the explanation attitude that we should be careful not to confuse them. In the process of constitution there is an interplay between them: the idea of truth, though not the truth itself or any truth-condition, is at work while constituting a logical connective and on the other hand logical connectives are used to grasp and represent the determinant, metaphysical principles which serve to determine the validity of an argument.

The mentioned conceptual distinction has not been observed in the model-theoretical semantics, for there the roles of logical connectives are exclusively determined by means of the (alleged) general conditions governing the truth. The logical connectives, there, are truth-functional. So that, for example, the validity of a determinant rule which is represented by the formula *ex falso* is already contained in the meaning of implication, and thus we are not able to discuss that what would differ if $\bot \rightarrow p$ would not hold. We are not able to discuss it because the validity of *ex falso* is already presupposed in the (truth-functional) definition of implication. Likewise, negation is defined, instead of as a relation among propositions (forming an intention based on the other),

---

formalism, but for Husserl the axioms would be the representation of apodictic intuitions, which accords his overall criticism against formalism.

3.3. PHENOMENOLOGY OF DIALOGICAL LOGIC 205

as representing the metaphysical principle of bivalence so that the meaning of negation as such is not separable from ex falso and law of excluded middle. Consequently the meaning of different connectives are confused; and this is an essential feature of two-valued truth-functional semantics that logical connectives are interdefinable, and even all can be reduced to one, which is very strange in respect to the intentional use of these connectives in reasoning and argumentation.

Proof interpretation and inferentialism are much better in this respect. They make us able to somehow separate the level of consequence-logic by assigning derivation rules, namely *introduction rules* and *elimination rules*, for any logical connective. The question that whether these rules are appropriate or not is another matter, but the point is that here we can grasp separately what follows from the meaning of a given connective. But still there are some problems. Beside some philosophical presuppositions that proof interpretation commits, which are not acceptable phenomenologically to which we will return in the following, the separation of the logical connectives each from the other and all from the structural rules are not pure. This is more evident in the case of negation which still presupposes the validity of ex falso. We may observe that proof-theoretical framework can not manage to appropriately distinguish between consequence level and truth level, for it still appeals to the notion of "proof" or inference, which are certainly connected to truth, to explain the meaning of the logical connectives and their role in a consequence relation.

The mentioned, distinction, however, is clearly observed in dialogical semantics, in its distinction between particle rules and structural rules and correspondingly the difference between play level and strategy level. The structural rules in dialogical logic correspond to what I called determinant rules and they determine the conditions of the valid argumentation. What follows from the meaning of a logical connective is the way of attacking and responding to it, regardless to that whether or not the proponent or the opponent can indeed perform such an attack or response. This latter should be determined by the structural rules. The role of the proponent is to defend a thesis and that of the opponent is to challenge it. The meaning of a given logical connective has nothing to do with these roles. So particle rules are *symmetric*. On the other hand, the validity of the argument, or what constitutes a proof as such, is dependent on those roles. So we have rules concerning what the rights and the obligations of the two parties are and how a dialogue should terminate. These rules also determine the result of the dialogue. So, the structural rules are in principle *asymmetric*. Then we have very nice criteria to distinguish between consequence level and truth level. From the phenomenological point of view it makes a lot of sense. As regards the truth, we deal with the clear and distinct judgment, so if the ego divides itself in two parties the one who is supposed to defend should commit to this idea of truth but the challenger, since it attempts to challenge and does not aim to establish a truth, is more tolerant in its acts. But in the consequence level, we have no obligation directed to truth then it should be neutral in respect with the two parties. So in one level we have rules governing how the connectives should be treated by the parties, and in the second level we have rules governing a dialogue as a whole and the specific roles that each party is supposed to perform.

In regard to the activity performed in the argumentation, now we are able to avoid a common confusion. From one side we have act of neutral derivation, or a mere operation on propositions, and from the other side the act of choosing the order of such derivations to construct a proof. In dialogical logic, in which we represent argumentation as a kind of competition, we call the first kind of activity *play level* and the second *strategy level*. So, to have a proof for a thesis is that the proponent has a winning strategy for it. Of course the play level is not responsible for truth or validity. It is possible that the proponent (or opponent) may win a play

3.3. PHENOMENOLOGY OF DIALOGICAL LOGIC

without this meaning that the thesis is valid (or not valid) since it is not *a casual* play which is determinant but to be able to win any play beginning by the very thesis regardless to whatever moves the adversary choses to do. Therefore, the notion of proof and that of logical truth is tied up with having a winning strategy. Then in another stage, the distinction between consequence logic and truth logic is reflected on the distinction between play level and strategy level.

3.3.3 Dialogical framework on the meaning of logical connectives

Now if we want to discuss a principle of logic, the dialogical framework provides us with a tool to analyze the effects of each connective or each metaphysical assumption at work. It is important to notice that the dialogical semantics by itself does not impose any particular logical system. It is enough to change some rules to obtain a variety of existing logics. And it is very interesting that in most cases what is needed is only to modify some structural rules; it means that most logics share the meaning of logical connectives; so that their discords can be rendered to the difference between structural rules. Such a phenomenon is called, after Došen who observed it in the proof-theoretic framework (Došen, 1980), *Došen's principle*. This is an obvious witness against the claim that each logic speaks about its own constants in the sense that a comparison is impossible. Such a claim of course is false for independent reasons, if we favor the phenomenological attitude; however it is interesting to notice that our dialogical approach clearly shows its deficient in a technical framework as it makes manifest the authenticity of the Došen Principle.

Accordingly we are able to consider some different logics, which their particle rules would differ from the standard ones, as having new constants not necessarily challenging the old ones. Such a phenomenon, namely introducing new constants by means of formulating new particle rules without altering the structural rules, is called by Rahman (Rahman and Keiff, 2005) *Girard's principle*.

The possibility to alter particle rules or add new ones, without changing the structural rules, makes us able to examine various alternative formulations for the logical connectives—and evaluate whether they are genuinely present in the act of reasoning and argumentation as such or are just of regional significance—thus study the consequence level of logic without this affecting or being affected by the studies pertained to the truth level—that is correlatively the studies on the basic metaphysical principles.

On the other hand, we can perform phenomenological investigations on the general principles governing the formal truth, namely the issues about structural rules, without they being confused by the issues belonging to the characteristics of the logical constants.

Therefore, the distinction between the structural rules and the particle rules can be seen as a realization of the phenomenological reduction: in order to grasp the essence of the logical connectives we put aside the features which are not essential to them, hence the particle rules. Likewise, in order to grasp the peculiarity of the formal argumentation we reduce it to its own characteristics disregarding the characteristics of the constants used within it, hence the structural rules.

The current project is a study on the meaning, then we will deal with the particle rules. One may apply the phenomenological method of *eidetic variation* to evaluate different and alternative structural rules, for example to show that the intuitionistic structural rule is the correct one not the classical structural rule. However, such a study belongs to the truth level; and I leave it in this project. Nevertheless, the method of eidetic variation can be used to work with the particle rules in order to chose the appropriate ones and thus explain the meaning of the logical connectives. Now, it means that in the dialogical framework, where the meanings of

3.3. PHENOMENOLOGY OF DIALOGICAL LOGIC

the logical connectives are to be analyzed in terms of the interaction between the two parties of the argumentation, we may approach to *seeing the essence* of the logical connectives by means of the variations applied to the patterns of those interactions, whereas the variations are eidetic because we have suspended other issues—including those related to truth—and our hypothetical agents are ideal ones. Therefore we have a framework to clarify and make precise the intentionality behind the logical constants, not just in a merely formal and mathematical but a properly phenomenological way. I will carry out this program to a certain extent in the last chapter.

3.3.4 Proof or evidence?

The difference between judgment and proposition is acknowledged in the contemporary studies on logic, after Frege until the recent developments in constructive type theory (CTT). So, the recognition of this distinction is not restricted to phenomenology. However, what can be conceived of from phenomenological investigations, which is absent, for example, in CTT, is the distinction between evidence and proof. According to phenomenology what is the ground of a judgment, or what makes a proposition true, is evidence, but according to CTT, and more generally to the proof interpretation of logic, it is proof which grounds a judgment and makes it different from a mere proposition. Such an attitude is not without shortcomings.

The difference between proof and evidence is not explicit in the dialogical framework. However, I think it provides suitable tools to grasp the distinction and observe its relevance to logical reasoning. In order to advance the phenomenological explanation of the dialogical notions of *play level* and *strategy level* we need to discuss, though briefly, the distinction between proof and evidence and explain why we should in general speak of evidence, rather than proof, while studying the judgment.

Proof is an ideal construction, while evidence can be ideal or not. Thus a proof once constructed remains forever; and this is explicit within intuitionism and also has been formulated by Kreisel (Kreisel, 1967, p. 159) as an axiom[19]:

$$(\forall n : N)(\forall m : N)(\Box_n A \rightarrow \Box_{n+m} A) \qquad (3.1)$$

It says if a judgment is asserted as true at the moment n this will remain true forever (remember that in the context of intuitionism truth is nothing but provedness). The above formula is acceptable for proof then it is true in the intended context namely mathematics in which the ground of our judgments are ideal proofs. However if we want to take this to explain a peculiarity of judgment as such, namely if we posit that judgment should depend on proof, then we face significant restrictions. It is obvious that a judgment which was once right can become wrong, and this is quit normal while speaking of the material state of affairs or mental states. Husserl, while explaining that the propositions as ideal forms of judgments are omnitemporal, declares:

> It should be noted that this omnitemporality [of proposition] does not simply include within itself the omnitemporality of validity [or truth]. We do not speak here of validity, of truth, but merely of objectivities of the understanding as suppositions and as possible, ideal-identical, intentional poles, which can be "realized" anew at any time in individual acts of judgment—precisely as suppositions; whether they are realized in the self-evidence of truth is another question. A judgment which was once true can cease to be true, like the proposition "The automobile is the fastest means of travel," which lost its validity in the age of the airplane. Nevertheless, it

---

[19]The mentioned one besides two others are called the axioms of the creating subject. Here I use the notation that Van Dalen and Van Atten use in their writings. See for further studies (van Atten, 2004)

can be constituted anew at any time as one and identical by any individual in the self-evidence of distinctness; and, as a supposition, it has its supertemporal, irreal identity.[20] (Husserl, 1973, p. 261)

What makes true a proposition such as "the sky is blue" is not an ideal proof but a piece of evidence which is material in its essence. Likewise what makes true, say, "I am sure" is not ideal nor material but a subjectively accessible mental state.

Beside the fact that proof is permanent but evidence is not, there is another feature in this distinction which is quite important. The realm of evidence as such is continuous while every proof is a distinct construction; and if we can speak of the realm of proof it would be discreet. Then even in a hypothetical case in which we ignore the change of the reality, still we are not allowed to replace evidence with proof.

So, what is the relation between proof and evidence? Obviously proof is a kind of evidence and not vice versa, but what kind? and what are the peculiarities of proof which differentiates it from other kinds of evidence? This questions is particularly important for us because in a significant part of logic we deal with proof. For, if the premise-judgments are grounded on evidence in general, what makes true a logical conclusion of these premises is a proof. Then

---

[20]"Es ist dabei zu beachten, daß diese Allzeitlichkeit nicht ohne weiteres in sich schließt Allzeitlichkeit der Geltung. Von der Geltung, der Wahrheit sprechen wir hier nicht, sondern bloß von den Verstandesgegenständlichkeiten als Vermeintheiten und möglichen ideal-identischen intentionalen Polen, die als dieselben allzeit wieder "realisiert" werden können in individuellen Urteilsakten — eben als Vermeintheiten; ob realisiert in der Evidenz der Wahrheit, ist eine andere Frage. Ein Urteil, das einmal wahr gewesen ist, kann aufhören wahr zu sein, wie etwa der Satz "das Auto ist das schnellste Verkehrsmittel" im Zeitalter der Flugzeuge seine Gültigkeit verliert. Gleichwohl kann er als dieser eine, identische von beliebigen Individuen allzeit in der Evidenz der Deutlichkeit wieder gebildet werden und hat als Vermeintheit seine überzeitliche, irreale Identität." (Husserl, 1939, p. 313)

in formal logic we are working with proofs and it is indispensable to have an idea of its relation to evidence in general.

As everywhere, we have the fundamental distinction between intention and fulfillment. Intention have various categorial forms of which propositional is of a particular importance for us. What fulfills a propositional intention we call evidence; by so fulfilling the ego performsan act of judging and as a product we have a judgment. Evidence stands in the side of reality (Wirklichkeit) so it is generally continuous, while intentions, and so propositions, are discrete entities. Then the distinctness of a judgment is due to its ideal form namely proposition. In most of cases the discreteness of reality is due to the objectifying and categorizing character of our intentions, so in most of cases while we express a judgment we can not separately express its evidence for if we have a case of self-evidence, evidence is exactly what manifests itself through the very act of judging, so what we have to express about the evidence of our judgment is but the judgment itself. It is in respect with the context that it would be made clear that a sign A is used to indicate a proposition or to serve as a judgment. To employ notions like $a{:}A$ to indicate a judgment while A stands for a proposition and a for what makes this proposition true, which is used in CTT, is in someway misleading. Because such a notation implies that prior to the judgment we have a proposition A as distinct and determined and a piece of evidence a as distinct and determined. If we restrict ourselves not to judgment as such but to proved judgments, then we could speak about the distinct proof-object—as will be explained in the following. Nevertheless, in general we have a determined A (a distinct intention) and the cognition of a and its identification and distinction is performed through the very intention A. Otherwise we must have another intention toward what fulfills an intention and so ad infinitum which is absurd. So, the notation '$a{:}$' will be useful only if we speak about the proved judgments (for example in the context of mathematics) not judgment in general, then the

3.3. PHENOMENOLOGY OF DIALOGICAL LOGIC 213

approach could not be generalize to logic as such.

Statuses of evidence are modes of being. In other words corresponding to every ontological region we can have evidence with a specific nature, hence ideal, material, mental and so on. Accordingly a piece of evidence may or not be intersubjective. Namely it is not necessary for a judgment that its evidence would be intersubjectively accessible, for example consider the assertion "I am happy". In the case of the material, evidence would be intersubjectively accessible but not permanent. In the case of the ideal, say for the judgment "2 is greater than 1", evidence is both intersubjective and permanent. In proof we deal with this latter kind. In contrast to the formers in case of the ideal the evidence may itself be already constituted as distinct; I would say this is because of the peculiarity of the realm of the ideal, because the ideal can be constructed by means of the activity of the ego, at it is constructed as determined and distinct.

Another distinction which we have studied before is also relevant here: the distinction between *intuitive* intention and *signitive* intention. In the former case an intention is directed to its object as intuited, namely this is the intuition of an objectivity which grounds an intention, in the latter case an intention turns toward an objectivity only signitively, namely when the object itself is not presented in the intuition. Any signitive intention does ultimately rest on some intuitions. Accordingly, and on the other side, a fulfillment of an intention can be intuitive or signitive. In the case of the proposition, this introduces us to the important distinction between *direct evidence* and *indirect evidence*. We use the terms direct evidence for the intuitive fulfillment, and indirect evidence for the signitive fulfillment, in the case of the propositional intention. For instance consider the proposition "I am hungry", the intention at work here may be intuitive, namely I already have an intuition of a state in my inner perception so that my turning toward such a state is based on its presence. Then such an intention is fulfilled intuitively, namely the proposition "I am hungry" is made true by a direct piece of evidence. On the other hand there are intentions that are formed signitively. Such intentions may be afterward fulfilled intuitively but also they can be fulfilled not in pure intuition. Consider the proposition "John is hungry", here I have no intuition of John's inner state, however on the bases of my intuition of hunger and also empathy I can have such an intention. In this particular example a possible fulfillment of my intention would be signitive, namely my intention toward that objectivity is fulfilled by means of some observations hence through a chain of mediations. So, this is indirect evidence which would ground the judgment "John is hungry".

Therefore, from one side we have the distinction between ideal and non-ideal evidence and from the other side the distinction between direct and indirect evidence. From this point of view we have four possibilities. As an example of direct ideal evidence consider the evidence for the judgment "2 is greater than 1". As an indirect ideal evidence consider the evidence for the judgment "π is an irrational number". For non-ideals we can consider the above examples of hunger. Notice that the ideal is also intersubjective while the non-ideal may or may not be so.

Now we are able to define "proof". *Proof is a construction of the steps of ideal evidence, so that itself serves as an ideal, indirect evidence for a given judgment.* The paradigmatic example of proof is the mathematical proof. So, a proof is an ideal construction, intersubjectively accessible and based on some apodictic intuitions of the ideal, which as indirect evidence grounds a judgment which is thus considered as proved. In an extension of the meaning the term proof may be applied in the cases we have direct ideal evidence, but such a use of the term is not without problems, mainly because in the case of self-evidence, proof can not be considered as a separate entity or construction as usually expected from the term "proof".

Notice that in the case of signitive fulfillment the process of fulfillment itself can be intuitive.[21] Likewise, in the case of proof, the fact that a proof is constructed in the pure intuition does not mean that the result of this proof is itself intuitive. So, for example, in intuitionist mathematics there is no need that every mathematical object or mathematical judgment be intuitive, what is needed is that the construction should be intuitive.

3.4 Dummett and Verficationism

3.4.1 Dummett on the philosophical importance of intuitionism

If we refute the naive realism, as the phenomenology does, this will effect both the theory of meaning and the foundations of *a priori* sciences, including logic and mathematics. The transcendental phenomenology is not alone in rejecting the naive realism. Dummett is one of the famous writers of the last century who criticized realism. As it is well known he elaborated the philosophical implications of intuitionism. He says:

> Those who first clearly grasped that rejecting realism entailed rejecting classical logic were the intuitionists, constructivist mathematicians of the school of Brouwer. (Dummett, 1991, p. 9)

Heyting, a former student of Brouwer, formulated a logic to be suitable to be employed within the intuitionistic framework. This logic turned out to be apart from any many-valued interpretation. From the philosophical point of view, such a result is not surprising. Classical logic, a two-valued system, like any many-valued logic, is founded on an extensional and thus realistic account; and if we rule out the realistic presupposition, our logic would not be based on a

---
[21]See for explanation (Husserl, 2001b, p. 725).

many-valued interpretation. From the technical point of view, this result was shown in a rigorous way by Göodel (Gödel, 1932).

Therefore, intuitionistic logic can not be conceived of as based on a truth-functional definition of the logical connectives. Then, beside the philosophical considerations against the algebraic meaning-explanations of logical constants—the considerations which are discussed, above all, by Husserl—now there is a formal system which avoids any truth-functional meaning-explanation. Dummett, while explaining the philosophical importance of intuitionism, offers an alternative meaning theory. He rejects truth-conditionality and supports a kind of verificationism, or as one may say proof-conditional theory of meaning. He tries to use the insights and achievements of the intuitionism to endorse his own point of view. In this section I wish to show that, Dummett's approach, although acknowledged in his critical considerations, can not be admitted by the phenomenological method. The transcendental phenomenology admits the rejection of truth-conditional theory of meaning and likewise grants the philosophical importance of the intuitionism, but it does not admit that this would entail supporting a proof-conditional theory of meaning.

In the following, first I sketch out Dummett's argument in favor of verificationism, as far as it appeals to the intuitionistic approach, and then examine it from the phenomenological point of view.

3.4.2 From the verificationist theory of truth to the verificationist theory of meaning

In his *Elements of Intuitionism*, Dummett says:

> On the classical or platonistic understanding of a universal statement like $\forall x \neg A(x)$, however, the truth of the statement in no way depends upon the existence of a uniform reason, or finitely many uniform reasons, why it should hold.

3.4. DUMMETT AND VERFICATIONISM

...

> From an intuitionistic standpoint, such a conception of truth [evidence-transcendent] for mathematical statements is an illusion. (Dummett, 2000, pp. 3–4)

The rejection of the evidence-transcendent, and in the case of mathematics, proof-transcendent conception of truth is completely in accordance with the phenomenological attitude. Therefore in this regard, intuitionistic mathematics is of a philosophical significance, for as Dummett rightly emphasizes:

> On an intuitionistic view, on the other hand, the only thing which can make a mathematical statement true is a proof of the kind we can give: not, indeed, a proof in a formal system, but an intuitively acceptable proof, that is, a certain kind of *mental* construction. (Dummett, 2000, p. 5)

Dummett's argument for the verificationist theory of truth, and his preference of intuitioinistic mathematics over the classical for this reason, is a quite adequate argument from the phenomenological point of view. However, he moves to a verificationistic theory of meaning.

> From an intuitionistic standpoint, therefore, an understanding of a mathematical statement consists in the capacity to recognize a proof of it when presented with one; and the truth of such a statement can consist only in the existence of such a proof. (Dummett, 2000, p. 4)

What Dummett has so far said was about the indispensability of the capacity to have a proof in regard to the truth of a statement. Namely he was arguing that "the possibility of being true of a mathematical statement consists in the capacity to recognize a proof of it". But now he uses "understanding" instead of something like "the possibility of truth".

> From a classical or platonistic standpoint, the understanding of a mathematical statement consists in a grasp of what it is for that statement to be true, where truth may attach to it even when we have no means of recognizing the fact; such an understanding therefore transcends anything which we actually learn to do when we learn the use of mathematical statements. (Dummett, 2000, p. 4)

Dummett attributes a truth conditional theory of meaning to the classical standpoint, then apparently keeps its core thesis but tries to refine the notion of "truth" used in it, so instead of truth conditional notion of meaning he is going to support a proof conditional or verificationist theory of meaning:

> The meaning of each constant is to be given by specifying, for any sentence in which that constant is the main operator, what is to count as a proof of that sentence, it being assumed that we already know what is to count as a proof of any of the constituents. (Dummett, 2000, p. 8)

It seems that Dummett's argument goes as follows

1. For the classical viewpoint, meaning is determined in terms of truth conditions.

2. For the intuitionist, truth should be understood in terms of proof.

3. So from the intuitionistic point of view, meaning is determined in terms of proof conditions.

Such an argument is inconclusive. Because we may reject what the classical viewpoint apparently claims in the first item. So if we want to replace truth with proof or provability we are not obliged to

3.4. DUMMETT AND VERFICATIONISM

restate the first item now in terms of proof. We have independent reasons to reject what the classical position claims the item one. Then at least Dummett's argument is based on an unevaluated presupposition and gives no positive reason for what it claims, except one based on a Wittgensteinian type of appealing to the notion of learning, which I will discuss below. On the other hand there are independent reasons to reject proof conditional theory of meaning as well as there are reasons to reject the truth conditional theory of meaning.

Dummett says:

> The classical meaning of the logical constants is given by truth-tables, which in turn depend on every statement's having a determinate truth-value. (Dummett, 2000, p. 7)

And then, having assumed such a background, he goes to replace truth there with proof, and concludes that the meaning of the logical constants should be grounded in proof conditions instead of truth conditions.

But what if even the meanings of the logical constants of the classical logic are not truth-functional? In dialogical logic we are able to show that by changing a structural rue we will obtain classical logic instead of intuitionistic logic, while all the particle rules remain the same. It shows that truth-functional approach is not intrinsic to the system of classical logic; rather truth-functional approach is a philosophical attitude which may evaluate classical logic as more appropriate than the other systems. We are to evaluate intuitionist logic as adequate in contrast to classical logic, however in order to do the evaluation we suppose that there is a common ground of invariant meanings which makes the evaluation possible. Husserl or Brouwer could say that for example the law of excluded middle is not valid regarding the proper meanings of negation and disjunction—not regarding some chosen meanings which would fit the rejection of that law. The fact that a term is used in a wrong or invalid phrase does not change its meaning, contrariwise it is the identity of the meaning which makes it possible to judge some proposition as valid and some as not. Husserl and Brouwer did not say that LEM is meaningless or obscure, they say that it is not valid. It means that they took it as clearly meaningful so that the meaning of negation or disjunction is as it should be but the judgment is not true in general, namely the classical logic do wrong in drawing some judgment from this very meaning.

Indeed, we can understand a statement without prima facia recognizing a capacity in ourselves to check its truth. If we recognize that there is no such a capacity available we put such a statement aside as absurd or as insignificant, but yet we are able to do that exactly because we *understand* that statement. A statement like "there is an odd perfect number", though there is seemingly no capacity to recognize a proof for it, is completely understandable; it is meaningful. And it is different from "not horse and human is" which is meaningless. The latter is not understandable not because we have no capacity to check its truth but because it is just meaningless. And on the other hand it is possible that a statement is meaningful but we can not speak about its truth or falsity. Of course one may use the word "meaningful" in a specific way to point out the significance of an element in regard to a given domain, so that the statements to which we can not attribute truth or falsity be insignificant or "meaningless" ; however, such a conception of meaning is not the same as what is concerned with in meaning theory or in any theory of language. I have distinguished such a notion by the term "significance" (sec. 1.4.1).

Dummett rejects a naive realism in regard to truth but in this way he appeals to a kind of use-theory of truth and accordingly a use-theory of meaning which does not necessarily follow from the former.

3.4. DUMMETT AND VERFICATIONISM

As discussed above, Dummett assumes that the argument for the essentiallity of the proof in regard to the truth counts also as an argument for the essentiallity of the notion of proof for the meaning. Why should we assume that "the classical meanings of '∨' and '∃' have no clear sense."? Contrariwise they do have clear sense, and indeed the meaning is the same for the intuitionistic point of view, except that the intuitionist does not admit a general way that the classical approach looks at what should be considered as the proof of the statement containing those constants. Dummett argues that classical treatment of the proofs concerning '∨' and '∃' is improper. That is right. But suddenly he says that the classical meanings of those constant are improper, without giving any reason to this transition from proof (truth) to meaning.

The claim that the meaning of logical connectives are the same for both classical logic and intuitionistic logic is not just a prescription rather it is clearly demonstrated by the dialogical semantics in which the particle rules for both logics are the same and the structural rules are different. Accordingly it is shown that the meaning of any logical connective is independent of the others as well as it is independent of the whole structure. This point is in accordance with the motivation of the intuitionistic approach. Dummett himself rightly says:

> Intuitionism rejects such a holistic view [i.e. the formalist view] of mathematics: for it, just as for Frege, each mathematical statement has a completely specific meaning of its own, a meaning which renders it capable of those applications which are made of it, but which is independent of any supplementary empirical hypotheses upon which such applications may hang. ... from an intuitionistic standpoint, ... it is the meaning of a statement which determines what is the appropriate notion of truth for it, in what we may take its being true to consist. (Dummett, 2000, pp. 2–3)

This remark stands against the use-theory of meaning which would ultimately lead to a holistic view. I would rephrase the remark as follows: " it is the meaning of a statement which determines what is the appropriate manner to use it, namely to prove it as true or to disprove it, not that the use determines the meaning". This also includes the meaning of logical connectives. Then, every logical connective has its independent meaning whether used in a true statement or in a false one or in a statement which is taken to be necessary or contingent.

Dummett asks:

> By admitting as legitimate the classical explanations of the sentential operators and the quantifiers, one does not rule out of order the intuitionistic explanations of them; but it is not very clear why, save as an exercise, one should then be interested in using sentences whose logical constants are to be understood according to their intuitionistic senses. (Dummett, 2000, p. 250)

But in fact, the meanings of the constants remain as it must be but some judgments concerning them are changed; more importantly this is the philosophical attitude toward the nature of mathematics which is different so that some basic posits and general rules may differ. For example consider the negation, assume that both of us, me an intuitionist and a classicist, do know what negation is, but he claims that $\neg\neg p \to p$ is true regardless to the content of p, namely its truth *follows* from the meaning of $\neg$, but I say that no, it *does not follow*. Here we both assume that we are talking about the same meaning. Otherwise there is no disagreement. But there is indeed a disagreement between intuitionism and other attitudes toward mathematics (or logic). Then I am not "interested in using sentences whose logical constants are to be understood according to their intuitionistic senses" rather I use the same logical constants but avoid making judgments about them which are not intuitionistically admissible but only classically assumed to be true.

3.4. DUMMETT AND VERFICATIONISM

Dummett says:

> The intuitionist is restricted in what he will acknowledge as an effective method of finding a number of a given kind precisely because he cannot grasp the senses which the classical mathematician wants to attach to the logical constants; but, if someone thinks that he can grasp those senses, there is no reason why he should regard as having any special interest those methods which can be recognized as effective even by someone who cannot. (Dummett, 2000, p. 251)

There *is* a reason. The reason is that only those methods are *reliable*, only those methods genuinely present us to the truth (non-effective methods are in fact inconclusive) so I restrict myself to those methods in order to achieve the reliable knowledge. Of course such methods exist in the classical mathematics but since there are also other (unreliable) methods the whole system is impure. It is not the case that the classical mathematics or logic is not understandable for me, rather they are undesirably mixed of truth and non-truth whereas I am only interested in the genuine truth. I disagree with them about the notion of truth so what they accept as true is not always true for me, so I should do a basic revision. The disagreement is really about the notion of truth not about the meaning of mathematical or logical expressions. In the formal level, the disagreement is represented in some axioms or rules, namely there are axioms (like LEM) or rules (like a specific form of *reductio ad absurdum*) that they admit as correct and applicable every where, but I do not accept, so I do not rely in their theory because they freely use these improper devices.

3.4.3 Dummett's argument as it relies on the notion of learning

Dummett's appeal to the notion of proof conditions is tightly linked with the use theoretic account. He says:

> An understanding of a sentence may now be taken to consist in a knowledge of the condition under which a statement has been conclusively established to be true; and no difficulty can any longer arise over what such knowledge consists in, since the relevant condition is, necessarily, one which we are capable of recognizing; in fact, meaning is now being explained directly in terms of what we actually learn to do when we learn to use the sentences of our language. (Dummett, 2000, p. 260)

Here Dummett describes the truth conditional approach then modifies it. But in essence a provability-conditional theory is the same as truth-conditional except that the notion of truth employed is modified. Some phrases above Dummett rightly indicates a main problem of the truth conditional theory, i.e. the question that what would count as to know such conditions. In order to avoid circularity there must be some sentences such that knowledge about them should be implicit. Then he describes this implicit knowledge as follows:

> there will be no difficulty in saying what is required for someone to know the condition for a sentence to be true, provided that the condition in question is one which he is capable of recognizing as obtaining whenever it in fact obtains, namely that he should, whenever the condition obtains, accept the sentence as true. (Dummett, 2000, p. 259)

It says that the meaning of some sentences are given in terms of their truth conditions which can be linguistically expressible and finally there are some sentences whose truth conditions are known implicitly and the implicit knowledge serves as the ground for their meanings. But now since we reach a subjective sphere why still we speak about the being able to recognize a condition once obtained instead of possessing merely an intention? It is like to say that I

3.4. DUMMETT AND VERFICATIONISM

know what blue means because I am able to recognize blueness of something if it is blue (and if I am in the position to see it). Why not to say the reverse: I am able to recognize blueness of something if it is blue (and if I am in the position to see it) for I know what blue means? Of course phenomenology says that this latter is the case. Dummett prefers to speak of recognition as the ground for meaning instead of meaning as the ground for recognition. However, as I have pointed out in the previous chapters, meaning in general is independent of recognition. A meaning as such does not necessarily depend on the possibility of intuition or recognition, but in the genetic level it has a peculiar relation with the intuition.

Dummett makes his use of Wittgesteinian ideas explicit:

> An argument of this kind is based upon a fundamental principle, which may be stated briefly, in Wittgensteinian terms, as the principle that a grasp of the meaning of an expression must be exhaustively manifested by the use of that expression. That is, as already observed, the understanding of an expression cannot, in general, be taken to consist in the ability to give a verbal explanation of it, and hence must constitute implicit knowledge of its contribution to determining the condition for the truth of a sentence in which it occurs; and an ascription of implicit knowledge must always be explainable in terms of what counts as a manifestation of that knowledge, namely the possession of some practical capacity. When it is a knowledge of the meaning of a word that is in question, then the practical capacity which constitutes that knowledge must itself be a linguistic ability, an ability to use or react to sentences containing the word in some manner that can, ultimately, be specified without appeal to any semantic notions assumed as already understood. (Dummett, 2000, p. 260)

Then Dummett can be criticized in the same way that we tried to do against Wittgensteinien ideas. The point is that Dummett attributes such an attitude to intuitionism and he thinks that perhaps at least one of the differences of intuitionism with classical attitude is that intuitionism is more fit to the Wittgensteinian idea. But I have shown that at least there is a tension between Wittgensteinian theory of meaning and that of phenomenology and yet at least Wittgenstein's theory is not the only alternative for the descriptive (or model theoretic) attitude toward meaning. If intuitionism is akin to phenomenology, which in the present project I take as true, and if there is a tension between Wittgensteinien theory of meaning and the phenomenological theory which is discussed in the first chapter and if the main reason to attribute a kind of provability-conditional theory of meaning to intuitionism is to appeal to Wittgesteinian ideas, then we should say that there is no reason to consider intuitionism as being akin to provability-conditional theory of meaning. Namely Dummett's argument is inconclusive in claiming that intuitionism is related to a verificationistic theory of meaning. I emphasize again that it is obvious that intuitionism supports the verificationistic theory of truth.

Following the above quotation, Dummett mentions three platonistic objection to that line of thought and responds them. But he does not consider an intuitionistic objection which may appeal to the phenomenology to reject the use theory of meaning and endorses an intentional theory. Instead, Dummett benefits from intuitionistic ideas to confront the mentioned objections but strangely he uses such ideas to show that we have no way to get rid of the dominance of language. Strangely enough he does not mention that for Brouwer the practice of mathematics is languageless so to appeal to intuitionism to support a linguistic-use theory of the meaning of mathematical constants is highly paradoxical. Dummett states a Wittgensteinian idea and mentions some platonistic objections to it but perhaps an intuitionist would precede to object.

3.4. DUMMETT AND VERFICATIONISM

3.4.4 Transcendental phenomenology on realism vs anti-realism

Realism is rejected by the phenomenology since it is a clear example of radicalizing the natural attitude. However this does not mean that phenomenology admits "anti-realism", for most of the stances referred to by the term anti-realism fall into the natural attitude as well. I will discuss this issue in the following.

Dummett's refusal of realism guides him to take a particular account for the logical constants.

> To show that a form of argument is valid in the semantic sense requires some kind of reasoning. If the reasoning itself involves the form of argument to be justified, then, most philosophers suppose, the justification is in effect a *petitio principii*; if it does not, then it amounts to a derivation of that form of argument from others, and its formulation in semantic terms is not significant. On this View, we have no option but to accept as valid certain basic forms of argument without further grounds for doing so. Since few would want to claim that we thereby evince direct insights into the structure of reality, the only alternative account is that, by treating such forms of argument as valid, we impose on the logical constants those meanings which we choose, and are free, to assign to them. We thus come back once more to the theory of meaning, but in a manner which strikingly reveals the different approaches that philosophers and logicians customarily take to concepts they share. (Dummett, 1991, p. 23)

The alternative of which Dummett speaks here is not admissible for the phenomenology; in the final analysis it is shown to be fallen under the natural attitude and is not free from the unevaluated prejudices of realism. On the other hand "direct insights into the structure of reality" is acknowledged by phenomenology without it being committed to realism. I shall explain this in the following.

However, Husserl speaks of *The Correlation of Theory of Meaning and Formal Ontology*[22] (Husserl, 2008, p. 50); and it is by no means the case that "we impose on the logical constants those meanings which we choose", rather we investigate those meanings as genuinely objective as far as related to the genuine studies on logic as theory of science. Husserl says:

> It will be shown that belonging to the essence of objectivity as such are systems of laws that, extending over all possible determinate, given objectivities, must only be ranked, not with a particular science, but with the science of science in general, and finally that, together with those grounded in the idea of meaning, all these laws form an intrinsically unitary theory of science, namely in that both are connected essentially by *a priori* bonds of thought, thus are linked to one another by bonds of thought that can be made evidently perspicuous.[23] (Husserl, 2008, p. 51)

Here, Husserl explicitly talks about the issue that Dummett mentions:

---

[22] Die Korrelation von Bedeutungslehre und formaler Ontologie (Husserl, 1985, p. 51)

[23] Wir werden sehen, dass dem wirklich so ist, es wird gezeigt werden, dass in der Tat zum Wesen der Gegenständlichkeit als solcher Gesetzlichkeiten gehören, die über alle möglichen bestimmten und gegebenen Gegenständlichkeiten sich erstreckend, nicht zu einer Einzelwissenschaft, sondern nur zur Wissenschaft von Wissenschaft überhaupt gerechnet werden müssen; und endlich, dass alle diese Gesetze mit den in der Idee der Bedeutung gründenden zusammen eine innerlich einheitliche Wissenschaftslehre bilden, indem nämlich die einen und die anderen Gesetze wesentlich zusammenhängen, durch apriorische, also durch evident einsichtig zu machende Gedankenbande miteinander verknüpft sind. (Husserl, 1985, p. 52)

3.4. DUMMETT AND VERFICATIONISM

> [P]recisely in connection with this correlation, just as each law of inference can be viewed as a law of validity for propositions of a certain form, so, in an obvious conversion, it can be viewed as a law for the obtaining and not obtaining of states of affairs.[24] (Husserl, 2008, p. 52)

Dummett says "On this View, we have no option but to accept as valid certain basic forms of argument without further grounds for doing so". We shall explicitly reject this. Then the consequence that Dummett draws, i.e. the freedom in choosing the meaning of logical constants, is groundless. Indeed the basic forms of valid argument are correlatively the basic laws of formal ontology and they are subject to certain, transcendental investigations.

Therefore:

> In the pure theory of meaning, we have a field of laws and theories that, in virtue of the correlation discussed, deals with objects as it does with meanings and extends over all particular sciences in the same manner.[25] (Husserl, 2008, p. 52)

Then, we are not "free" while speaking of the meanings of the logical connectives, for they are supposed to be capable to represent the objective connectivities of the formal ontology. Of course there would be a question that how such dependent meanings can possibly be objective so that they can be genuinely intuited instead of merely

---

[24][D]ass eben mit Beziehung auf diese Korrelation jedes Schlussgesetz wie als Geltungsgesetz für Sätze gewisser Form so in selbstverständlicher Umwendung als Gesetz für Bestehen und Nicht-Bestehen von Sachverhalten angesehen werden kann. (Husserl, 1985, p. 53)

[25]In der reinen Bedeutungslehre haben wir ein Gebiet von Gesetzen und Theorien, das vermöge der besprochenen Korrelation wie über Bedeutungen so auch und in gleicher, über alle Einzelwissenschaften hinausreichender Weise über Gegenstände handelt. (Husserl, 1985, p. 53)

being chosen. This is the question concerning categorial intuition about which we have discussed in the previous chapter (sec. 2.4).

The method of transcendental phenomenology does not fall under a bare realism nor a bare anti-realism. As discussed in details in sec. 2.7, the phenomenological method acknowledges both the essential role of construction and also the capability of being objectively intuited in regard to the ideal entities. Therefore, the phenomenological theory of meaning, as well as its philosophy of logic and mathematics, sees no dichotomy between realism and constructivism; and from an external point of view, it may seem to possess some elements of both trends.[26]

3.4.5 Against linguisticism

Dummett says that learning is essential for it is not the speaker who makes a relation between a sign and a meaning (Dummett, 1994, p. 50). But for Husserl this is the speaker who makes such a relation—of course in an original manner; and we may assume that afterwards such a relation turns out to be a rule to be observed by the speaker and his/her community.

> der Sprechende "erzeugt" die Mitteilung, d.i. den Ausdruck als Einheit von sprachlichem Zeichen und Bedeutung. (Husserl, 2005, p. 28)
> the speaker "generates" the message, i.e. the expression as a unit of the linguistic signs and the meaning.

However, as Husserl says, expression is itself two-sided as produced and as understood and expression is in its essence intersubjective. Any question related to expression in the production side has a counterpart in the receiving part. Then if I have access to a meaning and I indicate it by a sign so that I express something I suppose

---

[26]Gödel also share this view by his modified platonism. A good discussion of this position and its critiques to the other well-known positions can be found in (Tieszen, 2005, pp. 167–176).

3.4. DUMMETT AND VERFICATIONISM

that every hearer has, or rather is able to have, access to that very meaning. Intersubjectivity of meanings, as ideal entities, is not constituted within a language; rather language already presupposes such an intersubjectivity.

Husserl indicates that not every thinking is linguistic (*sprachliches*) (Husserl, 2005, p. 21). Nevertheless:

> Nicht das sprachlose Denken, sondern das ausdrückliche Denken wird für uns das Erste sein müssen. (Husserl, 2005, p. 22)
>
> Not the speechless thinking, but the expressional thinking must be the first for us.

The expressional thinking does not presuppose a language rather it establishes that. For this reason the adjective expressional, *ausdrückliche*, works better than the adjective linguistic. Of course we have conceptual cognitions that presuppose a discourse of concepts, thus a language, but it is not exclusively this case which contrasts non-conceptual cognition. In a deeper level, in some sense deeper than both non-conceptual, sensual cognition and conceptual, linguistic cognition, we have expressional cognition in which categories and concepts are constituted, hence a meaning system which makes possible a system of signs as a language, so that such a level is non-linguistic in one sense and linguistic in another sense.

In any case, to appeal to the linguistic issues to develop a theory of meaning is strongly rejected. What is vital for the theory of meaning is the notion of *expression*, which either in the primordial or non-primordial cases is prior to any ordinary language.[27]

3.4.6 Summary of the section

From our point of view intuitionistic logic is philosophically more important than the classical logic. Classical logic, as well as some

---

[27]We have discussed the relation between expression and different accounts of language in sec. 1.4.3.

other logical systems, may be of some special interests in respect with algebraic or computational issues. However, as far as the pure logic is concerned, the superiority is with intuitionistic logic. In this respect we are in agreement with Dummett. That is why we had to clear our discords with this philosopher in order to make it clear that although we are in agreement with him on this central thesis, we do not admit all the reasons he gives for and all the consequences he draws from this thesis. Dummett's main reason is that intuitionism does not commit [a naive] realism as classical attitude does. This is acknowledged by the phenomenology.

However, Dummett goes to endorse anti-realism, and through this way argues for a kind of linguistic pragmatism and verificationism. Neither anti-realism nor linguisticism is acceptable from the phenomenological point of view. Transcendental idealism is *not* a kind of anti-realism and refutes that there would be such a dichotomy between realism and anti-realism. A close study would show that to appeal to the notion of language-learning in order to explain the constitution of meaning would fall under naturalism, and thus from the phenomenological view point it is as far from the proper philosophical (or transcendental) attitude as realism is.

Another important root of the disagreement with Dummett is that he considers the notion of assertion as paradigmatic for the meaning theory. In this respect he is in the same way with most of his opponents. For example John McDowell (Auxier and Hahn, 2007, pp. 351–366) takes this for granted but tries to show that assertion (expression of a thought in his terminology—of course "expression of a thought" is different from "assertion" but, as Dummett points out in his reply, McDowell does not observe this distinction) depends on the idea of truth prior than that of justification, then meaning should be conceived of truth-conditionality. Then it is worth emphasizing that here while criticizing some ideas of Dummett we by no means side ourselves with his opponents in the analytic philosophy. We can figure out that Dummett keeps assertion or "establishment a sentence

3.4. DUMMETT AND VERFICATIONISM 233

to be true" as central, and meaning is connected to assertion, but, since he grounds assertion on proof instead of the truth, he connects meaning to proof. From our point of view, as explained so far, assertion or its ideal form, namely proposition, is not the ultimate form of the meaningfulness. Meaning has various forms, one of which is proposition. Different forms of meaning, corresponding to different types of expression have their own constitution (and have peculiar relations among themselves) and in each case may be brought about either primordially or non-primordially.

Dummett's appeal to the ordinary use of language and language learning is of a Wittgensteinian background and falls under the same criticism which we discussed in chapter 1; however Dummett even does not maintain a central insight in Wittgenstein's discussions: that is assertion is not the ultimate form of expression and it is not the only original form of the meaningfulness. Dummett leaves the one good lesson that might be learned from Wittgenstein and still speaks about justification and proof while trying to explain the constitution of expression and meaning.[28] For phenomenology, each categorial form and each type of expression has its own peculiarity as just said. Moreover these are ideal forms which may be represented by verbal signs which constitute a language. In each case meaning is constitutionally prior to an indicating sign. Also in the case of assertion, it is to utter a judgment while judgment itself is brought about thanks to an experience of coincidence between an intention and its fulfillment. So, Dummett's view that "judgment is an interiorisation of an assertion" (Auxier and Hahn, 2007, p. 369) is wrong. Assertion as a linguistic act depends on judgment. One should not understand this in the way that in the expression of a judgment the ego expresses the *act* of judgment as if the judgment itself is not expressional. Judgment itself is expressional

---

[28]Such an attitude has been diagnosed, as a wrong way to do studies on language, by Bühler and he calls such an inappropriate attitude *epistemologism* (Bühler, 2011, p. 216).

(notice the difference between expressional and linguistic which has been explained in the previous section). Indeed the (inner) act of judgment and the outer act of assertion may coincide, but what grounds an assertion is its representation of a judgment (as a product not as an act) not the other way around.

A very important disagreement between our philosophical method and that of Dummett is that for him the meaning does originally emerge within the ordinary language. This is an obvious discord with the transcendental phenomenology and also with intuitionism which one of its great observations is the languageless of mathematics. Dummett simply does not care about this very profound intuitionistic principle while he argues for the philosophical significance of intuitionism. Dummett argues for the advantages of the so-called intuitionistic meaning, discussing the importance of proof or demonstration there, which he connects with the process of language learning; whereas such a connection is by no means essential. As Sundholm and van Atten remark:

> Intuitionistically, to give a demonstration of a mathematical theorem is not to produce a certain linguistic object, but to produce a mental mathematical construction (or a method to obtain one, which method is of course also a mental construction) that makes the corresponding proposition true. (Sundholm and van Atten, 2008)

Finally, although Dummett argues for the philosophical relevance of the intuitionistic logic, and its basic argument would be acknowledged by the phenomenology, his attitude has important philosophical tensions with intuitionism, and with the transendetal phenomenology as well. Accordingly, his theory of meaning, although uses the insights of intuitionism, has no intrinsic connection with the intuitionistic approach and indeed should be rejected by it. Despite certain principal accordances with Dummett's ideas, the transcendental phenomenology disproves some of his other ideas,

3.5. LOGICAL MONISM

most importantly his proof-theoretical approach toward meaning and his linguistic account of meaning-constitution.[29]

3.5 Dialogical Framework and Logical Monism

Although, dialogical logic can be interpreted in a way akin to proof interpretation, it indeed is detached from both model-theory and proof-theory. The fact that dialogical framework is not model-theoretic should be obvious from the outset. Our discussion in regard to the distinction between proof and evidence may help to clarify the fact that dialogical approach need not to appeal to the notion of proof in the play level. Therefore, the meaning of the logical constants, which are to be represented by means of particle rules, are explainable without referring to proof conditions. As, some scholars say, dialogical framework can be seen as a kind of the so-called game-theoretical semantics. Marion says:

> The leading ideas of any game semantics are to define logical particles neither in terms of truth conditions, nor in terms of introduction (and elimination rules), but in terms of rules for games between two persons, a proponent and an opponent, and to define logical validity in terms of the existence of a winning strategy for the proponent. (Marion, 2012, p. 143)

However, in the present project I prefer not to use the term "game semantics", both for that I focus only on dialogical approach and do not deal with other semantics which are called in this manner, and for that I want to avoid any presupposition which is imposed by

---

[29]We have here only mentioned some objections against Dummett's ideas in regard to meaning; to study some phenomenological critiques to Dummett's overall approach, see (Hopp, 2009) and (Puhl and Rinofner-Kreidl, 1998).

using the term "game"—since here we will not able to go through a phenomenological discussion in order to make our conception of game clear. Nevertheless, what Marion says is indeed correct for the dialogical approach. In dialogical logic logical particles are defined in terms of rules for interactions. These rules are indeed eidetic rules, and the interaction need not to be considered a social one. The interaction, which is the basic notion of dialogical framework, can be seen as an inner dialogue—between the ego and its modification. Therefore, the dialogical framework is the most appropriate to carry out the investigation on the intentionality behind logical processes. Accordingly, the dialogical logic is in essential accordance with the transcendental phenomenology and may contribute to this latter in order to advance in logical meaning-explanations and to develop a formal apophantics.

As a powerful tool, dialogical framework may cover various logical systems, that is formulate different structural rules and incorporate various particle rules. This feature can be seen as advantageous for any project in logical pluralism. However, what is more important is that, the framework that dialogical semantics provides makes us able to *evaluate* different logical issues, so we may approach to a unified logic by examining different alternatives and incorporating any phenomenologically admissible feature. Such a project may proceed in both structural and particle levels. Structural level involves the studies on characteristics of the genuine argumentation and basic metaphysical principles, while particle level involves the studies on the meaning of the logical entities. Since our project is about the meaning and intentionality, in the next chapter I will discuss the phenomenological meaning of some logical connectives and will try to show the contribution of the dialogical framework to make precise the intentionality behind them and to formulate them accordingly.

Chapter 4

Dialogical Apophantics: Formal Analyses

4.1 Ideas toward a Unified Logic

The phenomenological method leads to a normative standpoint toward logic. Although there are various logical systems possible according to the formal criteria like consistency, such a pluralism should not be admitted for the philosophical logic, or what is called, in phenomenology, pure logic. Pure logic as taken to be the general theory of science or the science of general forms of reasoning must be universal. It is correlated with formal ontology. Although it is possible that specific theories of reasoning are developed correlating different ontological regions, all of them should be based on pure logic and are posterior to it. Then different logical systems, while being treated phenomenologically, either should be evaluated in order to reach a united pure logic or should be determined as being local so that their limits and specific notions would be the subject of semi-material studies. That is, we are not saying that pluralism about logic is wrong any way, rather we say that pluralism in regard to pure logic is wrong, and the pluralism about different semi-material logics should rest on a unified pure logic.

The achievements of the dialogical semantics, which are almost shared by the proof-theoretical framework, enable us to approach the idea of a unified logic. In the previous chapter I mentioned Došen's principle. According to this principles different logics can be formed by changing the structural rules while the particle rules remain the same. That is, particle rules can be determined independently of the whole system we are to develop. What should be done then, for the mentioned phenomenological project, is to investigate the nature of logical connectives and determine the right rule for each of them. The second phase would be to evaluate alternative structural rules and determine which of them genuinely represent the process of reasoning. While for the second phase I take it already admitted that the phenomenology would support the intuitionistic structural rules, I think it is worthwhile to do an outline of the first phase here both in order to clarify the nature of logical constant and explain their meaning and also to develop the formal relations, which would be extensions of the intuitionistic logic, to approach the realization of the pure formal logic.

I begin with the observation that for most of the logical systems it is true that each of them contains some genuine insights about the logical relations and, disregard their probably inadequate extension from a strict phenomenological point of view, they formulated some basic notions which would otherwise remain ambiguous. Then the first task of unifying logics is to recognize these insights. For example suppose logic A treats implication in a way which is phenomenologically admissible. Logic B has different treatment with implication which is also admissible. Now on the basis of a phenomenological investigation, *eidetic variation* precisely speaking, it should be determined whether we have really two different connectives which are usually refereed to by a same term "implication" or we have a single connective which shows different behaviors due to other features of each of logics at hand. Now if we really have two connectives they have to be able to be incorporated in a same logic.

4.1. IDEAS TOWARD A UNIFIED LOGIC 239

This logic would not be equal to A+B[1], but the insight behind it would gather the insights behind logics A and B.

4.1.1 Some significant insights from different logics

In the following, I mention some logical features which are formulated within different logics. Here I do not want to argue for the acceptability of these features, according to phenomenology. I just want to bring them as examples to show then how these features can be gathered in a same logic and how the intuition behind them can be made clear and distinct within the dialogical framework.

Intuitionistic Logic. According to this logic negation is a case itself has its own content, namely when we assert a negative judgment it should mean that we have an experience which is by its own content negative. This experience according to intuitionism is the experience of something leading to absurdity. So, on such an account of negation, there would be no general justification for the principle of excluded middle, for there is no guaranty that for anything we either have experience of it or experience of its absurdity. However, *ex falso* holds[2] , since if p and not p both would occur it would mean that absurdity would occur and it is in the meaning of absurdity to trivialize every thing. If *ex falso* and LEM both were valid then the elimination of double negation would be valid, but now there is no justification for it.

According to phenomenology this observation of intuitionism is true and there is a genuine conception of negation on the basis of which LEM and the elimination of double negation are *not*

---

[1]More precisely, it would not be equal to A+B+a bridge axiom.

[2]Here I consider Heyting's system while speaking of intuitionistic logic. There are certain arguments against the validity of *ex falso* with respect to Brouwer's attitude (van Atten, 2009). About *ex falso* and whether or not it represents a genuine intuition, so that it can be incorporated in a phenomenologically admitted logic, I will discuss in sec. 4.2.4.

generally valid. Since I am only to provide examples to depart toward demonstrating how unifying these insights is possible I do not intent to explain the accuracy of these insights here. However, in this particular case, namely about the accuracy of the intutionistic treatment with negation according to phenomenology, one may find good arguments in (Lohmar, 2004) and (Lohmar, 1992). Concerning the the intrinsic affinity of intuitionism with phenomenology we may refer to (van Atten, 2010).

Classical Logic. However, the classical observation of the validity of LEM is not totally groundless. If we define negation as assertion of falsity, then if we take falsity simply as not being true, then the true claim that every judgment is either true or false would result in the claim that for every judgment either it or its negation is true. We can recognize that here negation does not need its own experience, it is defined simply as the lack of truth experience. So, we should admit that LEM is valid and in fact we use it widely in our arguments. However, the conception of negation which satisfies LEM is not in the position to satisfy *ex falso*, for here no notion of absurdity is at work. Although LEM itself is valid, to take it as concerning the very connective which validates *ex falso* has resulted in unjustifiable claims such as the elimination of double negation. So, we want to take the intuition behind negation as validating LEM, but we want to avoid other relations within classical logic. That is why intuitionistic logic is, for phenomenology, acceptable as a whole though it may lack some genuine connectives and relations, but classical logic is not.

The solution within a unified logic is to recognize two different negations so that keep in distance the intentions behind what are ambiguously represented by the same term "not". I will explain this in the next section.

Logic of strict implication. There is a broad conception of implication which covers different intentions. However, in some contexts we speak about implication in a narrower sense so that to

4.1. IDEAS TOWARD A UNIFIED LOGIC

use other kinds of implication in lieu of it may cause some errors. We may roughly define strict implication as the case in which the consequence does not need something more than the antecedent to be true. Such a kind of implication was introduced by Hugh MacColl then formulated, as it is well-known, by C. I. Lewis. We may bring it in a unified logic if we can grasp its logical features in their purity. The intuition behind the strict implication can be explained as follows. While according to the general implication we have:

$$((\psi_1 \wedge \psi_2 \wedge ... \wedge \psi_n) \to \phi) \to (((\psi_1 \wedge \psi_2 \wedge ... \wedge \psi_{n-1}) \to (\psi_n \to \phi)) \quad (4.1)$$

this is not generally valid for the strict implication because it is possible that ϕ is not strictly implied by ψ_n, that is other premises would be needed.

Then, while p and q strictly imply p, it can not be inferred that p strictly implies that q strictly implies p. And although p and "p strictly implies q" both together strictly imply q, it is not generally justified that p would strictly imply that "p strictly implies q" implies q. Let's represent the strict implication by the symbol $\prec$, then the following formulas are *not* valid:

$$p \prec (q \prec p) \quad (4.2)$$

$$p \prec ((p \prec q) \prec q) \quad (4.3)$$

The rejection of some of the above formulas is the motivation behind some relevance logics. Our motivation here is close to that of Logic **E** but not Logic **R**, for in the latter the second formula above is taken to be valid. The fact that other relations of Logic **E** may be erroneous, especially that its treatment with negation is not acceptable, is another phenomenon. We are not to accept Logic **E** as a whole but we adopt its conception of implication as *a* genuine kind of implication, while dismissing that of Logic **R**. So our unified logic should not contain the Logic **E** as a fragment. However, on the basis of a distinct conception of the strict implication we may put it together with the general implication in a same logic, in which we would also have:

$$(p \prec q) \prec (p \to q) \tag{4.4}$$

Our implications would not collapse to each other, and no undesirable feature would arise. This is to be shown in the next section.

Paraconsistent Logic. Paraconsistent logic challenges the principle of explosion, that from contradiction, or from a absurdity in a precise reading, anything is entailed. This is not necessarily to say that the expulsion is undesirable because contradiction really exists. That is one trend, but the main claim is to make reasoning possible even if contradiction occurs in somewhere. The principle of expulsion, or *ex falso* partly relates to the concept of contradiction and thus negation and partly relates to the concepts of "entailment". I have said above, and I will show it in details, that there is a negation which does not validate *ex falso*. However, our both implications satisfy *ex falso* for the case of the strong negation. Nevertheless, still there is a genuine intuition of a kind of entailment or implication, which is widely used in the scientific reasoning, which says that the antecedent *effectively* implies the consequence. The intention here is something more than a logical consequence, though it must be properly represented by logic, in that here a kind of internal relation between antecedent and the consequence is supposed. For such an implication *ex falso* should not be valid, for the absurdity could not effectively result in something. To formulate such a conception of implication is a problematic task which I do not deal with here. However I hope my investigation on the kinds of negation and formulating a logic in which *ex falso* does not hold for a certain kind of contradiction could fulfill some genuine intentions behind paraconsistance approaches.

4.2 Negation

There are some different kinds of negation in the various logical systems. Each of these negations admits some of the features supposed to be relevant to the issue of negation and dismisses some other. From these features are the principle of explosion (*Ex Falso Sequitur Quodlibet*), the law of excluded middle (LEM), elimination of double negation, principle of non-contradiction and contraposition[3]; the accuracy of each of these features must be discussed in the first place through philosophical not merely formal studies. Of course, we may have more than one consistent system, just as a considerable number has been introduced in the last century; but what makes them different for us is the philosophical motivation they start with. Phenomenologically speaking, a system advances on the basis of primary, *apodictic* intuitions and although formal issues like that of consistence is important they are not originally responsible for the constitutional essence of a knowledge system. What is important is to be aware of this primary insights and to clarify them to a certain extent before entering in mere technical practices.

In the following, I shall first briefly argue that there is still a need to clarify the nature of negation and make its logical behavior precise . From the phenomenological point of view, the treatment with negation in the classical logic is not acceptable and that in the intuitionistic logic though quite admissible does not cover certain characteristics which are peculiarly reflected in some sort of negative judgments. This is, as I will argue for, because of the ambiguity of the notion of negation: even speaking of a one universal logical system we have more than one negation. Two different negations, irreducible to each other, are to be distinguished. Then I introduce

---

[3]By contraposition we exclusively mean (if p then q) $\Rightarrow$ (if not-q then not-p). The other way around, i.e. (if not-q then not-p) $\Rightarrow$ (if p then q), is a combination of contraposition and double negation elimination, then there is no need for it to be treated separately.

a logic which contains these two negations, formulating its semantic rules on the basis of standard dialogical semantics.

At the end of the section, I will come back to the philosophical discussions in order to analyze whether the introduced logic fulfills our primary intention and what tools the new formalization provides in order to carry out further clarifications around the topics related to negation and argumentation in general.

4.2.1 A review on the role of negative judgments

First of all, the concept of negation is connected to the notion of "false" as opposed to true. Consider the assertion "it is raining". Let's call this judgment p. Now, provided that it is a sunny day, one may say that p is not true. Here we have some untrue. There is a viewpoint saying that in such a case we have something true which is the negation of the judgment that has turned out to be false, namely not only p is false but also not-p is true. According to the strict realistic point of view, which is reflected in both Aristotelian and classical logic, this account is adequate everywhere and it represents the very nature of negation. But from the phenomenological point of view truth means truth-for-me ("me" as a transcendental ego, namely regardless to psychological desires and dispositions), or truth as ideally experiencable by a transcendental ego (who orients itself toward knowledge not that merely has some believes which happen to correspond some allegedly predetermined facts). So, not to experience the truth of p would not necessarily mean to experience another thing called not-p. This shows that the relation between falsity and negation is more delicate than what may seems at first glance. In regard to the more basic notion of truth we may understand falsity in either of these two ways:

1. We mean by truth and falsity both as experienced, which these experiences are mutually exclusive.

4.2. NEGATION

2. We mean only by truth it as experienced and have a more vast account of falsity while such an experience is lacked.

Regarding the fact that "false" is widely understood as the counterpart of true so that we safely use it in the absence of "true", namely without taking care about further concerns seeking some peculiar positive content, it is appropriate to take "falsity" as meaning the second item above. We will use this term in this sense in the rest of the present note.

If we fix the concept false as non-true, it is admitted that every judgment is either true or false. But as we said above this does not mean that there is always a truth about the object of judgment which is revealed while we decide the judgment as false. This is the point phenomenology diverges from realism. For example, while the realist says that either p or its negation is true regardless to the experience or possibility of experience of one of them, according to phenomenology this is not always true. For example, I cannot say that Goldbach's conjecture is true; but also I cannot say that its negation is true. This should not be interpreted as if there is a third value beside truth and falsity, rather it concerns the notion of negation at work and its relation to falsity.

In the following we briefly analyze the phenomenological account of negation, mostly based on Husserl's discussions in the sixth investigation of *Logical Investigations* (paragraphs 11 and 12) and in his *Experience and Judgment* (sections 71 and 72).

Negation and the act of judgment

Above all we should notice that the negative judgment is not primarily to express a peculiar objectivity, namely the alleged being-quality of falsity, or non-reality, rather it should first be understood as a certain modification of judgment in its plain and simple form. So, studies on negative judgments first of all are fallen under the more general investigations on judgment-modifications. Nevertheless, judgment as a whole is not the only place in which negation appears; we have also negation in the sub-propositional case, e.g. where we say "the sky is non-red."

However, our analysis here focuses on the judgmental negation. We are to investigate the negation as plays a peculiar role in the course of argumentation and as appears as a certain constant in the propositional level. We will refer to the issues about sub-propositional negation where it helps to make clear the nature of the judgmental negation.

In regard to the judgmental negation, two extreme views in opposition to each other can be recognized, one saying that the negative judgment is as basic as the positive judgment, the other saying that the judgmental negation is essentially reducible to a sub-propositional negation so that any negative judgment can be, in principle, replaceable by a positive judgment.

To explain the first view, it is helpful to recall the truth-functional semantics of classical logic. One may say that by a same function a negative judgment is converted to a positive and a positive to a negative; as if we have a dichotomy which exhaust all possibilities, and logic provide us with a tool, namely negation, to transit from one side to the other. Such a view is basically supported by the principle of excluded middle which is here applied to the positive/negative just as to the true/false distinction. This account and its ground to consider the negative judgment as basic as the positive are strongly rejected by Husserl and also as it well-known by Brouwer. We will briefly discuss Husserl's phenomenological analysis in the following section.

Are the negative and the positive twins?

Husserl, just from his very early works, explicitly criticizes the extensional attitude toward logic[4]. This is completely in accordance with the transcendental method which was to be developed later

---

[4]See (Husserl, 1994a) and (Husserl, 1994b)

4.2. NEGATION

on. The critics against extensionalism are dealt with in another occasion in this project. However, here we are to emphasize that any interpretation of logical topics which is based on an extensional account of it, though perhaps our mind are used to it and seems very natural, should be re-examined. The conception of negation as the *complementary* is one of these extensional interpretations. Putting such a conception aside[5], we should investigate the acts in which the (judgmental) negation is constituted. We should examine whether the constitution of the negative judgment is as basic as the assertive, positive judgment itself or if not what is the relation between the negative judgment with the more basic assertive one, and thus what role the constant negation plays in establishing such a relation.

In order to begin, let us assume the non-predicative intention toward "color of the sky", provided that we already have the intuition of a set of colors red, green and blue. On the basis of the former intention we may constitute another intention, which is propositional in its essence, that the sky is red. One may say that in this level we may constitute also an intention that the sky is non-red. This is the case of sub-propositional negation or conceptual negation in contrast to the judgmental negation. Here we focus on the latter and presuppose the constitution of the former. The question is how a judgment of the form not-p, like the sky is not red, or equivalently, it in not the case that the sky is red, is constituted. Is this a

---

[5]The fact that Husserl in the mentioned texts from 1891 seems to employ the notion of negation as the complementary(Husserl, 1994b, p. 109) is not in contrast with the phenomenological attempt to revise this notion. Husserl there does not aim to revise or justify logical relations as such, rather he wants to show that extentional attitude has no originality, let alone superiority, in comparison to the intensional interpretation, and for any alleged rule explainable by the extensional interpretation there is a conceptual explanation. So Husserl deliberately restricts himself to the formulas posited by Schröder whose attitude was the target of Husserl's criticism there. The need for revision and taking a normative standpoint toward logic is not excluded; in fact it is taken to be a very important task, as we have already discussed in the second chapter.

result of a direct and unfounded assertion? The answer obtained by phenomenological analysis is no. We may *infer* from the basic assertion "the sky *is* non-red" that "the sky *is not* red"—that such an inference is certain and indubitable does not refute the fact that it is different from a direct assertion. Also we may have thought that "the sky is red", but then recognize that it is not so and thus refuse our former judgment by asserting that "it is not the case that the sky is red". For this latter case we may either perceive that indeed the sky is blue, so the new judgment rejects the old one, or just recognize that what we considered as evidence for our judgment was not adequate. Imagine that we suddenly realize that we are in fact in a cave and the red we had perceived was the color of cave's ceiling, so we actually have not perceived the sky and are not able to judge about its color, then we assert that "it is not true that the sky is red".[6] In each case the negative judgment is brought forth as a sort of modification. Husserl declares:

> *[A]ffirmative and negative acts of position-taking, the position-taking of recognition and that of rejection, do not simply represent two coordinate "qualities,"* like red and blue in the sphere of color, for example, and that consequently the *expression "quality" in general is not suitable here. The act of negation of the ego consists in the exclusion of validity,* and the *secondary intentional character* [of negation] is already implicit in this expression.[7]

---

[6]The difference between the two cases here are an example of the difference between the strong negation and the weak negation which will be discussed in the next parts.

[7]"[D]ie bejahende und verneinende, anerkennende und verwerfende Stellungnahme nicht einfach zwei gleichstehende "Qualitäten" darstellen, wie etwa in der Farbensphäre rot und blau, daß somit die Rede von Qualität hier überhaupt nicht paßt. Das ichliche Negieren ist Außergeltungsetzen, und schon in diesem Ausdruck liegt der sekundäre intentionale Charakter angedeutet."(Husserl, 1939, p. 352)

4.2. NEGATION

And continues:

> It is a *basic error of traditional logic* that it established basic forms of the judgment without having clarified the sense in which one can speak of them and, above all, that it allowed negation (the negative categorical judgment) to appear among them. Opposed to this, it is necessary to emphasize that one cannot speak of a series of basic forms. There is *only one basic form*, which is *the simple* (positive, and not, for example, "recognitional") *categorical judgment:* "S is p."[8] (Husserl, 1973, p. 292)

Husserl then states that the act of refusal which can be expressed in relation to a judgment may itself enter in a new confirmation, then it is true that in logic and science we deal with this or that form of confirmation. Husserl explains in few lines, and then leaves, that the mode of rejecting may be included as a new moment in a positive act. The how of this inclusion is the theme we are to elaborate in this analysis. Is the moment of negation representable, in any case, in the side of the predicate? or it should be considered as a modality? or as a peculiar form of modification? The next section mainly discusses, and rejects, the first option. Indeed we are to continue the phenomenology of negation, and defend the thesis that considering it as judgment-modification we recognize that there are two kinds of negation each of which has its own modificational characteristics.

However, it is clear that every judgment is based on an act of fulfillment, so even in the case of rejection, it is due to "reversal

---

[8]"Es ist ein Grundfehler der traditionellen Logik, daß sie ohne Klärung des Sinnes, in dem von Grundformen des Urteils gesprochen werden kann, solche aufstellte und darunter vor allem die Negation (das negative kategorische Urteil) auftreten ließ. Demgegenüber ist zu betonen, daß nicht mehr von einer Reihe von Grundformen gesprochen werden darf. Es gibt nur eine Grundform, das ist das schlichte (positive, nicht etwa das "anerkennende") kat egoris ch e Ur teil "S ist p"."(ibid)

of attitude' '(Husserl, 1973, p. 293) that we have a new intention which is already fulfilled and makes us able to perform a judgment (positive but including a negative moment). The presupposition that considers negative and positive judgments as counterparts or as twines imposes unjustifiably an extensional view on judgments. One of more important results of this presupposition is to consider the function of negation as symmetric as if we will comeback to p by negating q if q is the negation of p. This notion must be put aside, because it is not granted that if we modify x, in some way, and reach y, then if we modify y in that way we will reach x. That whether this situation is true for negation or wrong, or admissible in some occasions, will be clear as we progress in our phenomenological investigation.

Are negative judgments avoidable?

We saw that from the phenomenological viewpoint the negative judgment is not constitutionally as basic as the plain judgment; then we reject that logic treats these forms as symmetric. On the other hand, there is another view which refutes any originality for negative judgments. Such a viewpoint has been introduced in the field of logic by Griss an intuitionist mathematician who tried to formalize a negation-less arithmetic. Here, our concern is not restricted to mathematics, so we prefer to give examples from empirical cases. Although our explanation then would not be faithful to Griss's motivation as such[9], it helps to grasp the insight behind such an attempt as it is supposed to concern the knowledge-oriented activities in general. We will try to explain the mentioned insight putting it in the phenomenological framework.

Suppose that we have an intention toward the sky being red; but

---

[9]Griss's objection against the use of negation in intuitionistic mathematics is based on the intuitionistic account of negation. However his attitude is inspiring in order to formulate and evaluate a possible thesis in regard to the negation as such, and thus to shed a light on some aspects of the nature of negation.

4.2. NEGATION

we experience the blueness of the sky so that our former intention is frustrated. This is an instance of a sub-propositional negation, an essential analysis of which can be found in Husserl (Husserl, 2001b, p. 702). As Husserl says there:

> If I think A to be red, when it shows itself to be 'in fact' green, an intention to red quarrels with an intention to green in this showing forth, i.e. in this application to intuition. Undeniably, however, this can only be the case because A has been identified in the two acts of signification and intuition. Were this not so, the intention would not relate to the intuition. The total intention points to an A which is red, and intuition reveals an A which is green. It is in the coincidence of meaning and intuition in their direction to an identical A, that the moments intended in union with A in the two cases, come into conflict....
>
> We may generally say: *An intention can only be frustrated in conflict in so far as it forms part of a wider intention whose completing part is fulfilled.*[10]

Our intention toward the sky being red happens to be frustrated

---

[10]Meine ich *A sei rot*, während es sich in "Wahrheit" als grün herausstellt, so streitet in diesem sich Herausstellen, d. h. in der Anmessung an die Anschauung, die Rotintention mit der Grünanschauung. Es ist aber unverkennbar, dass dergleichen nur möglich ist auf dem Grunde der Identifikation des A in den Akten der Signifikation und Intuition. Nur so kann die Intention an diese Anschauung überhaupt heran. Die Gesamtintention geht auf ein rot-5 seiendes A, und die Anschauung zeigt ein rot-seiendes A. Indem sich Bedeutung und Anschauung hinsichtlich der Richtung auf dasselbe A decken, treten allererst die beiderseits einheitlich mitgegebenen intentionalen Momente in Widerstreit, das vermeinte Rot (das vermeint ist als Rot des A) stimmt nicht zu dem erschauten Grün.

Allgemein werden wir danach sagen dürfen: *Eine Intention enttäuscht sich in der Weise des Widerstreites nur dadurch, daß sie ein Teil einer umfassenderen Intention ist, deren ergänzender Teil sich erfüllt.* Bei einfachen bzw. vereinzelten Akten ist also von Widerstreit keine mögliche Rede. (Husserl, 1968, §11)

only while our wider intention toward the sky is fulfilled and while we perceive the sky as blue. Then it can be said that our intuition is properly expressed in the judgment "the sky is blue". But at the same time as we know that blue is non-red, namely we know about the members of a set of predicates which are mutually exclusive (for the sake of simplicity suppose that our object is single colored), we may judge that "the sky is non-red". Now the Grissian thesis is that in each case we say that $\neg A(B)$, or that B is not A, we can replace it by B is non-A, while non-A is itself a given property. So, according to the mentioned phenomenological remark, in such a case always there is some C which is, according to its constitution, non-A, so that we could not say B is non-A unless that we already had judges "B is C", or $C(B)$.

In other words, the thesis is to say 1- any judgment level negation can be reduced to a predicative level negation and 2- a predicative level negation is not intuitive unless we had a positive predicative intuition, expression of which would make the negative judgment superfluous. The second part phenomenology supports, but let's take a closer look on the first part.[11]

---

[11]Although we are going to criticize this assumption here but it should be mentioned that it is not, also according to phenomenology, such an implausible position to take that is better to be excluded from the outset. In fact Husserl himself once found this assumption acceptable. He wrote in a text from 1891:

> All negative judgments can be transformed into equivalent affirmative judgments with negative matter. (Husserl, 1994b, p. 100)

Moreover, a cursory look on some passages from *LI* may makes it seem to the reader that Husserl considers negation as negative matter inside the judgment, but he there makes it explicit that he deliberately limited himself to the subpropositional negation and it is not a comprehensive study on the role of negation in regard to judgment (Husserl, 2001b, p. 706). Husserl certainly has changed his idea from what he stated at 1891; this is clear from his analyses in *EJ* as we saw in the previous section; otherwise there was no need to study negation under the topic of modification, for negation would not be modification of judgment but a change in its content.

4.2. NEGATION

A first set of doubts may arise as follows. Is every where we use a negative form a positive experience presupposed? Then what experience able to be expressed in a positive judgment leads me to say, e.g. "Centaur does not exist"? What about judgments of non-identity such as "Square is not circle"?

Griss' answer is nothing but appealing to the basic intuitionistic, or phenomenological, principles. We are allowed to judge about something if we have it in some sense in intuition or construct it in mind. As he wrote to Brouwer:

> Showing that something is not true, i.e. showing the incorrectness of a supposition is not an intuitively clear act. For it is impossible to have an intuitively clear concept of an assumption that later turns out to be even wrong. One must maintain the demand that only building things up from the foundations makes sense in intuitionistic mathematics.[12]

For example, the judgment "A square circle does not exist", for Griss, is rather meaningless, for what it possibly would be the intention behind such a proposition? While constructing the square we are aware of its being non-circle then the non-identity judgments are pointless, and for the very reason we have not an intuition of square circle to put it as a subject for proposition and predicate something to it. Arend Heyting explains Griss' point as follows:

> The question which Griss asks us can be formulated as follows: what is the exact sense of a proposition like: "A square circle does not exist"? Evidently, in order to understand this, we must first know what a square circle is: however, as such a thing does not exist, how can we have a clear notion of what it is? (Heyting, 1955)

---

[12] A letter from Griss to Brouwer dated 1941-04-19, published in (van Dalen, 2011, pp. 402-406).

Griss says that, in our words, an intention which cannot be appeared in a judgment except to be shown as impossible to be fulfilled could not be in fact formed from the outset. But one may ask, having accepted that there is no genuine intention behind "square circle", what about the judgment about Centaur? How we may judge about fictional creatures which are constituted out of some intentions though not based on perception? The answer would be that there is no need to negative judgments; it would suffice to say that "Centaur is non-real" which is a weaker expression of what is involved in its constitution namely that "Centaur is fictional" or "Centaur is mythical", provided that the mythical due to its constitution is non-real.

Then Griss claims that every where we speak of non-existence or non-identity it is either meaningless or we can safely replace it by speaking about some positive predicate or about *difference* while we can positively demonstrate this difference. For example Griss says:

> No real number a is known about which it has been proved that it cannot possibly be equal to 0 ($a \neq 0$), while at the same time it has not been proven that the number differs positively from 0 ($a \# 0$). (van Dalen, 2011, pp. 402-406)

(Brouwer introduced a counterexample for this, as discussed below.)

Husserl too speaks of negation as *distinction* in the case of experiences involved with negation (Husserl, 2001b, p. 701), but he makes it explicit, as we said before, at the end of that discussion that he had not dealt with judgment but a mere fragment of it (Husserl, 2001b, p. 706); then it is not right to assume that Husserl equates negation with distinction (or difference) or sees analysis of the negation in the property level sufficient.

For instance, instead of saying that 2 is not equal to 3 we can say that 2 is different from 3 by -1, and according to Griss we can do the like in every occasion.

4.2. NEGATION

Here arises an important objection, but we do not go through it but just mention. Heyting (Heyting, 1955) remarks if in order to avoid negation we are obliged in some occasions to consider some expressions as meaningless then we would have problems with formal reasoning for we must consider the contents. He notices that we might have to forget propositional logic at all. However, this only shows that Griss had to present precise criteria for meaningfulness which must be prior to his idea about replacing negation by positive assertion. He might still keep his attitude. As Heyting rightly stated then, "It is certainly too early to judge of the value of Griss' conception". But Griss' untimely death prevents us from being introduced to his further investigations. So, here I do not deal with a criticism which could be possibly solved with further works. I focus on the criticisms which aimed the essence of the thesis.

The second kind of objections to Griss's thesis is to show that there actually are negative judgments that are not replaceable by some positive ones. Brouwer (Brouwer, 1948) gives a rigorous mathematical example for this. Here we do not intend to go through mathematical arguments, but we can outline an informal argument which is inspired by Brouwer's formal one.

Recall the distinction between intention and fulfillment. Now it seems that Griss says everywhere we judge, an intention has been fulfilled, then instead of saying negative aspects we can assert a positive judgment without loosing any information. This is true in the case of predicative experience, as it is shown by Husserl. However we may judge about our intentions in higher levels. We may, on the basis of a fundamental intuition, say that a certain intention of ours is fulfillable though it has not been fulfilled yet. This is to reflect a characteristic of the transcendental ego which is not restricted in its current direction of intention. For example, it is a transcendental principle that any significant problem is solvable, no matter that I in the concrete life actually solve it or not. Now consider a yet undecided, but completely meaningful problem, like

Goldbach's conjecture. It is certain that this problem is NOT solved now. When it will be solved, the intention containing it will be fulfilled, but now there is a difference between unfulfilled intention and fulfilling act, whereas we do not have any positive information about this difference—otherwise we could solve the problem. It is acceptable that in the case of "Goldbach's conjecture is solved THEN (a date or a circumstances)," we can replace THEN with the term "NOT...now" in the former judgment and have a positive form. But in the present state we may judge about the state of the problem only in the negative way, that it is not solved.

Brouwer, in his paper *Essentially negative properties* (Brouwer, 1948) respond Griss' thesis and rejects it[13]. Brouwer defines a real number like r, in an exact manner which is determined by the free choice of the ego regarding how (or when) the problem is proved or disproved, otherwise it would be zero. There is basic evidence for that the number is not equal to zero (the problem is not insoluble), whereas there is no evidence to determine the difference of this real number with zero, namely it is true that $r \neq 0$ but not $r \# 0$ (provided that $\#$ is a relation giving the difference and can be defined in a positive way). So, Brouwer concludes: "Thus for real numbers the relation $\neq$ is an essentially negative relation." In other words a proposition like $\neg(a = b)$ can not be always reduced to a negationless proposition.

Consider that since this is an example from mathematics, and thus from the activity of the creating subject, we can not speak of concrete conditions to formulate a positive judgment. In the case of Goldbach's conjecture it is essentially unknown that when creating subject will reach a proof for or against it. This is essential for the spontaneity of the ego.

Then it seems that we would be able to avoid negative judgments only if, at least, we forget about our status as self-aware and autonomous and restrict ourselves, or more accurately our

---

[13]Reprinted in (Brouwer, 1975, pp. 478-479).

4.2. NEGATION

judgments, to non-reflective intuitions and judgments.

However, a basic insight of Griss' has remained unanswered; that is, if a plain form of negative judgment was false, then how we grasp a falsity and put it as something about which a judgment is possible while it is supposed not to be the case? Or more precisely, as negation in the intuitionistic framework is defined as implying absurdity how this notion of absurdity, which is absurd by its definition, can play a role in the constitution of logic and argumentation? This is what we are to deal with in section 4.2.4.

4.2.2 Kinds of propositional negations

In ordinary speech we may revise a previous claim of ours and negate it. This does not mean that necessarily we claim a new judgment. Assume, to use the example from page 246, Adam says "the sky is red", Eva says but it is the ceiling of the cave that you refer not the sky, Adam says "Oh I see, then it is not the case that the sky is red". Notice that in this case Adam merely retraces his commitment to the claim that p ("the sky is red"), and by asserting that "not" p he does not commit himself to the claim that "the sky is non-red". It may still be true that the sky is red, but in any case the evidence that Adam had for his claim does not work anymore. We can repeat this example with the Goldbach's conjecture: Adam claims "for all even numbers there are two prime numbers sum of which is equal to that even number", Eva says: "How do you claim that? This has been shown only for the numbers smaller than y." Here Adam may accept that not p merely in the sense that p is not true (or what is equal has not been experienced as true).

But, is the kind of use of "not", of negation, shown in the above examples the only one? At least in scientific arguments, we may *prove* not-p or posit it as experienced. For example Adam claims that π is an algebraic number, but here not only there is no proof to show that π is algebraic but also there is a proof to show that π is not algebraic. Then Adam is forced to negate his claim in a strong way (not in a weak way as in the case of Goldbach' conjecture). If we take this account of negation as the admitted one, then there would be no evidence for *principle of excluded middle* (PEM) in general. It is not the case that for every p we have either evidence for p or for not-p.[14]

In the above passages we repeatedly used the term "not", but as this is exactly what we are to clarify, we could not avoid ambiguity from the outset. After reaching a thesis and fixing a new terminology we may be clearer about the situations mentioned. However, we can see that some differences show themselves in the various use of not. We may guess that either it is not a single intention that is indicated by the expression "not" or the negative connective (notice that we are to speak merely about the judgmental negation), or there is a genuine intention behind negation and other uses in the ordinary speech are results of confusions. Let us begin with this latter option.

As we said before, according to phenomenology and also to intuitionism, truth of an assertion is experienced or experiencable by the transcendental ego, or the creating subject. Then if an assertion of the form not-p is to be a genuine assertion its truth should be the matter of an experience; it would not be enough not to experience p to assert not p. Then according to the above remark PEM would not hold in general.

According to phenomenology, every logical principle corresponds to a subjective law of intentionality and evidence, provided that this "subjective" here means transcendental subjective and does not refer to any psychological issue.[15] There is a subjective law of evidence

---

[14]I use the term evidence which represents a notion broader than proof. According to phenomenology the latter could not every where substitute the former. Proof can be considered as evidence, but we also have the case of self-evidence in which no separate proof is discernible (see sec. 3.3.4).

[15]For a detailed and well-explained discussion on this see (Lohmar, 2002a). We might just add to that discussion that while all logical principles are obtainable through *eidetic variation* and each is self-evident, only those which

4.2. NEGATION

to which corresponds the law of non-contradiction. The subjective law here says, to borrow the phrase from Lohmar (Lohmar, 2002a):

> No one is able to perform the judgments 'A' and 'not A' together with distinct evidence .

But there is no such a law for PEM, because it would say that for any proposition, the ego is able to judge either in favor of it or in contrast to it. As Lohmar states it:

> The claim that judgments are true or false 'as such' makes sense only under the implicit presupposition that every judgment can be brought to the evidence of self-givenness.

Then, as it is clear that such a presupposition is too strong and in fact not defendable in general, the principle of excluded middle, so far, cannot be accepted. This is completely in accordance with the intuitionistic rejection of the general validity of PEM, as discussed by Brouwer in (van Atten and Sundholm, 2016) and as he introduces a strong counterexample to PEM in (Brouwer, 1955) (reprinted in (Brouwer, 1975, pp. 551–4)).

So, do we have a single genuine intention constituting the notions of negation, so that some claims in which negation appears, e.g. PEM, are results of confusions?

But yet it seems that what is tried to be formulated in classical logic is not totally groundless. As our first examples show, we are able to withdraw an assertion and to express this intention in some way. It is no matter that we say "not" p or what else, but it would

---

are determined as axioms are self-evident in the *apodictic* way, and thus the judgments expressing them are constitutionally prior to others. Accordingly, a proper attempt to formalize logic must begin with investigating the apodictic evidence which form the axioms, which can be seen as rules in another manner of formalization. In the current discussion we want to extract the axiom(s) concerning negation from our apodictic evidence of the transcendental subjectivity as possessing the negation-intention(s).

satisfy PEM, for it would be clear that any intention is either fulfilled or not; we are to assert in relation to a proposition p, either it in the plain form or that it lacks evidence. We may observe that most of times confusion occurs between the two acts described, between two different intentions which are indicated by the term "not". But this does not necessarily mean that one of them is genuine and the other is ill-formed.

If we read not-A, instead of a judgment based on evidence contra A, a judgment asserting that A lacks evidence, then PEM turns out to be acceptable. But we are not to replace this latter reading of not in the place of the former one which is more familiar in the context of phenomenology and intuitionism.

Here may arise an objection: 'in the case of having no evidence for p, when we withdraw p and assert "not p", what would be the experience on the basis of which we may assert such "not p"?' This may be responded as follows. There is actually en experience which constitutes the truth of "not p" in this latter case but it is not an objectively-oriented experience, its an experience of our conscious state, which is possible thanks to the transcendental characteristic of the ego, that we may *judge* that we have no access to specific evidence. Then in the case of this latter "not p" we have a genuine judgment, though in respect to a judgment about the objectivity p it is only a modification.[16]

---

[16]It can be said that the weak negation is basically concerned with the ego's epistemic state. But this does not imply that weak negation then has nothing to do with the objectivity itself. Although it is in first place about the epistemic state, considering the transcendental correlation between consciousness and reality (*Wirklichkeit*), it would play an irreducible role in the course of reasoning and argumentation or knowledge oriented activities in general. In order to stress this moment we deliberately avoid using expression like "agent's epistemic state", which is not such a desirable expression within phenomenology—it indicates that agent itself is a real phenomena whose epistemic state is a separated piece of objectivity having some sort of psychological relation with the objectivity about which the argumentation goes. We prefer to speak of the ego's conscious state or the ego's awareness of its intentions being fulfilled or not. The first

4.2. NEGATION

We distinguish between the *strong negation*, which occurs in a judgment asserting that p is objectively rejected, and the *weak negation*, which occurs in a judgment asserting that there is no evidence for p. In the following we try to show how these two connectives can be incorporated in a single logic.

Before going to the formal investigation, we briefly discuss what features seemingly relevant to the negation are desirable and what features are not.

The principles of non-contradiction and contraposition are already sufficiently self-evident (however we will have a brief discussion for this in the following), also that p implies its double negation.

We can now see that the elimination of double negation is a result of a confusion between two accounts of negation. Consider the weak one: not-p means that there is no evidence for p, now why if we have no evidence to the claim that there is no evidence for p we would have evidence for p? Consider the strong one: not-p means that there is evidence against p. Now why what is against of against of p should indicate the truth of p? Putting the assumption of PEM aside, we know that if there is a truth about an objectivity this can be falsified in many ways, hence a truth and a plurality of falses.

The principle of excluded middle, as we discussed should not hold in general for the strong negation, but is one of the main features of the weak negation.

Intuitionistic logic (at least in the form introduced by Heyting) includes the principle of ex falso. The claim that if a contradiction occurs everything would be derivable is not as intuitive as non-contradiction and contraposition are. In fact there is an argument in the literature that intuitionistic logic, on the basis of the proof interpretation, should not accept *ex falso* (van Atten, 2009). However, this principle is quite relevant to the discussion on negation. As we just said, the strong negation occurs when a

---

person analysis has priority everywhere in our investigation.

peculiar experience about the objectivity related to p obtains. This is usually represented by $p \to \bot$, namely p would lead to absurdity. Then the concept of strong negation is explained on the basis of the concept of absurdity. Now it makes a crucial difference whether this absurd would have as its peculiar feature as destroying or trivializing everything, namely as satisfying $\bot \to p$, or not. The definition of the strong negation as $p \to \bot$ is quit reasonable for it would be absurd to do have evidence both for and against a proposition. However that whether $\bot \to p$ is genuinely true is controversial, and I am in agreement with van Atten that none of the existing arguments for ex falso is satisfying. However, I think that ex falso is the very peculiarity of absurdity, otherwise absurdity would have no characteristic and as a result negation would not be a specific modification but another arbitrary proposition. I will give an argument in favor of ex falso which will be based on the very meaning of absurdity and will in turn help better to understand this meaning. For now I just declare that ex falso holds for the strong negation but not for the weak negation. Ex falso does not hold for the weak negation because weak contradiction, that is the conjunction of a proposition with its weak negation, does not result in absurdity, that is, it is not absurd both to have a piece of evidence for a proposition and absence of such a piece of evidence although it would not be the case. Presence and absence are not *truly* opposing each the other; and to assume both accruing simultaneously does not involve an ontological conflict. This will be clear in the next section while discussing about the nature of absurdity.

To summarize, our two negation are to satisfy the following conditions:

Non-contradiction and contraposition hold for both.

Ex falso holds for the strong negation but PEM does not.

PEM holds for the weak negation but *ex falso* does not.

4.2. NEGATION

Elimination of double negation holds for neither.

We have said before why elimination of double negation is not acceptable not for weak negation nor for strong one. It can be shown that it would hold for any constant which satisfy both PEM and *ex falso*. But after separating these former features, elimination of double negation appears as an independent feature which can be decided in axiom level.

In the following sections, we will formalize a logical system, using dialogical semantics, which covers the mentioned characteristics.

Before that we briefly take a look on some important viewpoints which in some sense speak of a second negation or formulate a logical system containing two negations and explain why they do not cover what our intentions about judgmental negation involve.

4.2.3 To examine some suggestions for logics with two negations

There are some logics containing two negations, such as the approaches based on many valued semantics, Nelson's logic with strong negation, various Dual negations, Galois split negations, modal and epistemic approaches like that of von Wright to introduce a weak negation besides the classical one, (also works by ken kaneiwa (Kaneiwa, 2005a) (Kaneiwa, 2005b)). We may see some similarities in the motivation behind formulating those logics with our own. However a close study would make it clear that the basic philosophical attitude are not the same as ours, and thus the result, as we may expect, is different from what we are to reach. I omit to show the differences here in details. Once our suggested logic with two negations is introduced, a few page later, the formal differences with the mentioned systems can be simply checked. Nevertheless, regarding their philosophical relevance here I discuss many-valued approach and also I will examine some suggestions which have been introduced within the phenomenological and the intuitionistic methods.

Many-valued logics

There are some logics which contain two negations by means of introducing a third truth values. In order to abolish the PEM and the elimination of double negation one may consider to appeal to a third truth value so that the truth-functional character of negation changes in some desirable way. Such an approach is of course basically wrong from our phenomenological point of view. Because, as discussed in the previous chapter any model-theoretic, extensional or truth-functional account of the logical connectives could not give a proper explanation of their essence and accordingly plays no role in their constitution. However, it could be still fruitful to look at such attempts in order to examine if there are some insights behind them and if they notice some peculiarities of specific logical constance which remain neglected elsewhere.

One of the very nice works in this field is a short, and attractively written, paper by Varzi and Warglien (Varzi and Warglien, 2003). The article begins by the observation that:

> There are, naturally, two ways of thinking about ordinary sentential negation.

Which are:

> (i) as a form of complementation (the negation of a proposition p holds exactly in those situations in which p fails), and (ii) as an operation of reversal, or inversion (to deny that p is to say that things are *the other way around*).

Though the distinction itself is admitted, the insight behind the second conception is not so clear. What it means that "things are the other way around"? If by this we mean that not only p fails but also the objectivity about which p speaks *is* in a situation which rejects p (like the difference between, say, "the sky is not red" and "the absurdity of the sky being red is evident"), then the claim authors made immediately after the above passage is groundless:

4.2. NEGATION

These two conceptions are significantly different. But whereas a variety of techniques exist to model the intuition behind conception (i)—from Euler and Venn diagrams to Boolean algebras—conception (ii) has not been given comparable attention.

Given the interpretation we mentioned, it would not be true that the second one lacks significant attention. In fact it is this conception which is properly formalized within intuitionistic logic—the fact that intuitionistic logic seems not to be as popular as classical logic is irrelevant here. Furthermore, classical logic, though has benefited genetically from Euler and Venn diagrams and Boolean algebras, is not clearly focused on the conception (i). In fact we can recognize a confusion there. Then it is not true that conception (ii) has remained relatively untouched even in the popular trend, rather there it is treated improperly.

But if we read the phrase "things are the other way around" as speaking about exactly reversion, then the second conception is itself an erroneous one. Consider the proposition "the sky is red", how we can negate this proposition so that claims that the situation is the other way around? Should we say "the sky is blue" or the "the sky is green"? There is no significant conception of reversal here. If we mean only non-red or absurdity of sky's being red, then we would speak about intuitionistic relations. And I do not see any other way of understanding here. It is true that in some cases non-x is exactly the inversion of x, but it is not the case every where. For example consider the distinction between inner and outer. Of these two each has its own constitutional origin; it is not so that, say, outer is just a sign for what is not inner. Here we can say that inner is non-outer and outer is non-inner. Here reversal works, but is this the case every where? What is the inversion of "the number of planets of the solar system is seven", "the author of Organon is Aristotle" and so on? Or more problematically, what is the inversion of an arbitrary, meaningful but senseless, proposition, e.g. "happiness is red"? What would be a proposition that in response claims that the situation is the other way around? We are not saying that we can not judge in favor of the inversion here; we say that the inversion in this former case is completely meaningless. This is enough to show that this alleged kind of negation is not a genuine logical connective.

The strange thing is that in the approaches favored by the authors of the aforementioned paper, PEM also holds for the second conception of negation; in the system in which a third truth value is introduced the principle of excluded forth is taken to be the case instead of the principle of excluded third but in any case these are instances of PEM. Then the elimination of double negation (or n-negation in the case of n-valued) would hold in their logics.

The important point is that if we have a case for which inversion really holds then the elimination of double negation would be valid, but not as a logical relation, not as a characteristic of negation, but as a feature of the objective region at work. The fact that inner is non-outer and outer is non-inner would give that non-non-inner is inner; and if we strongly negate the strong negation of "It is in inner perception that time is given" then we would have that "It is in inner perception that time is given". But here we presuppose that time is given in either way, otherwise elimination of double negation would not hold.

The authors rightly speak of the confusion between two notions of negation. But instead of tracking the root of confusion, two-valued-ness is taken to be responsible and the disambiguation is sought for in many-valued-ness.

The authors try to widen the possibilities of "inversion" by analyzing many-valued logics, but the point is that the observation concerning "inversion" is wrong just from the outset, thus all widenings are pointless in respect to the notion of negation as such.

It is not surprising that the logic that the authors introduce at the end differs significantly from what we want and declared in the previous section.

4.2. NEGATION

Within the phenomenology and the intuitionism

In these two approaches which are our favored ones in this project, one can find some ideas about distinguishing two kinds of negation. We are to show that those ideas have not been elaborated sufficiently; and not only a logical system dealing with two negations has not been introduced but also the informal discussions, though very insightful, have not been very clear in order to distinguish two genuinely different intentions which are usually indicated by a same term "not" in the judgment level.

Husserl, as we said before, had a revisionist standpoint toward logic but he had not set for himself the task for developing a formal system of logic. Some may think that the lack of direct attacks to certain formal relations of classical logic and mathematics in Husserl's work, though there are a lot to their overall attitude, implies that Husserl only aimed at the interpretation of logic and mathematics not their content. This is certainly not true, and as far as concerns some particular theses of the contents of these sciences we may refer to certain phenomelogists after Husserl who examines the validity of those theses applying the method of phenomenology. As a distinguished example from contemporary scholars we should mention Dieter Lohmar who, among other things, has investigated the relation between Husserlian theory of categorial acts and the constitution of logical connectives. Lohmar (Lohmar, 1992), investigating the notion of negation according to phenomenology, speaks of a distinction between two kinds of negation, but briefly and without using the idea for further developments.

Having shown that elimination of double negation is not acceptable according to phenomenology, Lohmar claims that in a certain sense we could speak of validity of this relation:

> An appropriate solution would be the terminological distinction between the two forms of negation, which occur in the double negation. If we call the "simple" negation, which is the reaction occurring through the

not-experience of a plain intention, $not_1(-_1)$, and in contrast, the sign for the second, subjectively directed, negation $not_2(-_2)$, we can formalize:

$$-_2(-_1 A) \Rightarrow A \ ^{17}$$

From this, Lohmar concludes that the validity of the elimination of double negation in the traditional logic is groundless. However, Lohmar does not treat his idea concerning the two forms of negation independently and brings it in only in few lines within the discussion about double negation.

In the intuitionistic context, it is not uncommon to speak of weak negation but not as a genuine logical connective; and what we are here to formulate as the weak negation is different from what is called so in some intuitionistic works.

We may begin by the definition as Van Atten (van Atten, 2014) states it:

> One speaks of the "weak negation" of A to express that so far no proof of A has been found. This excludes neither finding a proof of A nor finding a proof of $\neg A$ later.

This is in fact based on the distinction which Brouwer (Brouwer, 1955)[18] made between four cases:

1. A has been proved to be true.

---

[17] Eine angemessenere Lösung wäre die terminologische Unterscheidung der beiden Formen der Negation, die in der doppelten Negation vorkommen. Nennen wir die 'einfache' Negation, die als Reaktion auf durchstreichende Nichterfüllung einer schlichten Intention erfolgt $nicht_1(-_1)$ und im Gegensatz dazu das Zeichen für die auch subjektiv gerichtete zweite Verneinung $nicht_2(-_2)$ können wir formalisieren:

$$-_2(-_1 A) \Rightarrow A$$

[18] Reprinted in (Brouwer, 1975, pp. 551-4).

4.2. NEGATION

2. A has been proved to be absurd.

3. Neither of the above cases has happened, but an algorithm is known to prove either A or that it is absurd.

4. Neither A nor that A is absurd has been proved and there is no algorithm known to decide this.

The case 3 is reducible to one of the former ones; then there are three cases. These three cases have been made explicit also by Heyting. However they do not consider this latter case as a kind of negation. The point made by Brouwer and Heyting can be considered as an informal explanation without trying to provide a ground for further formalizations on the basis of the hitherto introduced formal relations. Indeed they don't want to define a weak negation on the basis of strong negation. Heyting (Heyting, 1930), while speaking of the case different from both p and $\neg p$, which he called p', declares that this p' is not able to be definitely formalized. Of course we are going to somewhat disagree with Heyting, defending a thesis that there is a formalization which covers p'. However, we admit Heyting's observation that p' cannot be formalized based on the already existing notions.

Some may interpret the above distinction by recalling a notion of "to know" or "to be constructed" as a predicate, or operation, separated from the judgment itself, as if the case 1 says kA, the case 2 $k\neg A$ and the case 4 $\neg k\neg A \wedge \neg kA$. This is an interpretation, and I leave here whether it is adequate or not, but I argue that given this interpretation there is still room for defining a weak negation; and further formalization on the basis of the mentioned interpretation could not handle what we described in the previous section as required from the formalization of weak negation. The difference between the weak negation as we have described it and the case 4 above would be clearer once we have introduced the formal definition and the logic containing the new connective.

This latter interpretation is linked to what are known after Kreisel as the axioms of the creating subject (Kreisel, 1967, p. 159). He there introduces a novel notion to intuitionism:

$$\Sigma \vdash_m A$$

Here Σ stands for a creating subject; and the expression means "the [thinking] subject Σ has evidence for asserting A at stage m". The axioms that Kreisel posits using this notion are:

(i) $\Sigma \vdash_m A$ is decidable for each given Σ, m and A.

(ii) If an assertion A is true then for each Σ it would be absurd that it be absurd to exist a m in which Σ has evidence for A; and for each Σ if there would be a m in which Σ has evidence for A then A is the case.

Sometimes, the second axiom is stated in a strong form which says if A is true then there exists a m in which Σ has evidence for m. However, we are to deal here with the first axiom which afterwards formulated in this form:

$$\forall x (\Sigma \vdash_x A \vee \neg \Sigma \vdash_x A)$$

Van Dalen and Van Atten use the following notation in their writings and we will follow them:

$$(\forall n : N)(\Box_n A \vee \neg \Box_n A)$$

Since this is similar to PEM, it may come to mind that the weak negation we are seeking for is this $\neg \Box_n A$. This formula though correct in its own context namely in mathematics, seems not to be valid outside that context e.g. in empirical discourses. And in general the intention behind $\neg \Box_n A$ would not satisfy what we require from weak negation.

Let us notice that if $\neg$ is to mean the intuitionistic (i.e. strong) negation, as it is supposed to be, then $\neg \Box_n A$ would not mean that

4.2. NEGATION

the ego at the stage, or date, n has no evidence for A, rather it means that it is *absurd* that at the stage n the ego has evidence for A. Then would the axiom (i) be acceptable? There is an intuition behind PEM which we are to formulate by introducing the weak negation. This intuition says that for any proposition the ego either has evidence in order to judge in favor of it or has not. But the axiom (i) says another thing. It claims for any proposition at any given stage the ego either has evidence in order to judge in favor of it or it would be absurd to have such evidence then. The axiom (i) could not provide us with a weak negation but it itself needs to be reinterpreted on the basis of weak negation to survive as justifiable. Consider the case of Goldbach's conjecture. In our current state we have no evidence for it nor for its being absurd (hence the rejection of PEM for strong negation). But is this true that since we have no evidence for it if we would have, it would result in absurdity? In fact this is a very strange claim. One may think that Kreisel by saying that $\Sigma \vdash_m A$ is decidable had an informal notion of weak negation in mind. This is a transcendental principle that the ego is able to decide whether or not he has access to a fulfilling act for any particular intention (including the propositional). This "not" here is weak negation. Strong negation is reserved to cases in which for an intention there is evidence for its unfulfillablity. That is why PEM is not generally valid for strong negation. Now we can see that the weak negation cannot be defined by reducing it to the strong negation.[19]

---

[19]It may be tried to save the mentioned axiom by appealing to the following interpretation. For any propositional intention A, the ego may constitute another intention toward A being fulfilled in the particular moment n. If this latter intention is fulfilled we would have the judgment $\Box_n A$. Now if it is not fulfilled, since the moment n has passed, it would be impossible to have evidence in the moment n and we would have $\neg \Box_n A$. Then for any moment in time a given proposition A either has evidence or it would be impossible for it to have evidence in that moment. But this is nothing but to apply PEM to the continuum of time which is not intuitionistically acceptable. Such an interpretation relies on a presupposition which considers a moment as a

There is another point about the axioms of the creating subject which concerns our discussion. In the standard reading of those axioms we also have:

$$(\forall n : N)(\forall m : N)(\Box_n A \to \Box_{n+m} A)$$

It says if a judgment asserted as true at the moment n, it will remain true. Let us call this feature omnitemporality. As far as concerns to mathematics this is true, but as we want to speak also about empirical facts this is not correct. We have discuss this matter in sec. 3.3.4.

This issue is also important here because one may think that our definition of the weak negation is temporal and a genuine judgment should be omnitemporal. Having a version of $\neg\Box_n A$ in mind, one may think that weak negation is to say "there is no evidence for A till now". We say that this "till now" is redundant. Just as we do not say "I have evidence for A till now". If there is an assumption of a peculiar mode of temporality for judgments and it goes without saying, this is the case for the weak-negative judgments too. Also it is true that when we weakly negate an assertion we may be still in search to find evidence, and there is apparently no such a thing in positive or strongly negative assertions, but this does not mean that our judgment here is not a genuine judgment, rather it only means that our judgment here is less objectively informative than positive or strongly negative judgments.

Therefore, there is yet no sufficiently developed conception of the weak negation in the context of phenomenology and intuitionism. The few suggestions are not able to be considered as complete and precise. Departing from phenomenological observations we are going to explain the strong negation and then the weak negation in

---

district point. As if "now" is the point in which it is determined that either x happens or fails. This conception of time is not acceptable as it is shown by the phenomenological investigations carried out by Husserl. Then such an interpretation would stand both against intuitionism and the established phenomenological achievements.

4.2. NEGATION

comparison to it, in a way that neither of these two is reducible to the other. We will show that the logical relations to be obtained do not violate the intuitionisic logic (as formulated by Heyting), rather they may provide a tool to extend this latter to be more appropriate to be applied in the variety of argumentations. We will come back to discuss some of these possibilities in the section 4.2.6.

4.2.4 Strong negation; and the case of *ex falso*

We take the idea that negation can be taken as an implication, and absurdity as a proposition, as our point of departure. Then, indicating the rules which reflect the peculiar content of this proposition, we aim to determine the function of negation in the argumentation. Namely we take $\neg p$ as equal to $p \to \bot$, then state our thesis to express the rule corresponding to the proposition $\bot$. The idea that $\bot$ can be considered as a proposition is very crucial for our thesis; it is not merely to use a more convenient way of formalization. Since we want to use the proposition $\bot$ as basic out of which negation is constituted, it is not appropriate to read $\bot$ as contradiction.

If one reads $\bot$ as contradiction and then equates $\neg p$ with $p \to \bot$, this would result in circularity, for not-p would mean that p leads in contradiction and contradiction would mean to have both q and not-q together.

Then, contradiction is posterior to negation. Negation should be so interpreted that in $p \to \bot$, $\bot$ represents a concept other than contradiction, a concept more fundamental than both contradiction and the strong negation. Instead of contradiction, I think, it is appropriate to speak of absurdity as a primitive notion which serves as a ground for the concept of (strong) negation. However, absurdity itself is not intuitive. I am to defend that absurdity may be considered as a proposition whose constitution has been prior to any negative proposition. I will explain the meaning of this proposition in the next section. The fundamental intuitive proposition in which absurdity appears and is logically characterized is:

$$(p \vee \bot) \to p \tag{4.5}$$

That is if between absurdity and an arbitrary proposition one is true, it is that other proposition which is true.

This proposition, the absurdum, may directly be substituted in the general intuition that $p \to p$, then we have this a priori truth that $\bot \to \bot$; that is, absurdity is absurd, or, following Parmenides, it is not the case that nothingness holds.[20]

Now this proposition, the absurdum, can enter as a part in further categorial acts in order to construct other propositions, e.g. propositions of the form $p \to \bot$. This latter proposition may find its proper intuition; and, to be sure, this does not entail that p or $\bot$ must themselves be objects of intuition.

Therefore, although absurdity is not intuitive and it will never be intuited (for $\bot \to \bot$ is intuitive), it is a basic notion. Since this notion itself is not given in an intuition it seems that there is an open debate on whether its categorial form is proposition or concept or some other form of meaning. However, we would argue for it being a proposition, namely that we have also a meaning of negation (among some meanings which are ambiguously covered by the word negation in the ordinary language) based on a notion which is originally of the categorial form of proposition and in this form it has a significant place in the construction of propositions and argumentations.

---

[20]This is from a sentence from the second fragment of Parmenides's poem which has been translated in these forms:

- that it is and that it is not possible for it not to be (McKirahan, 2011)

- 'exists' and 'it is not possible not to exist,' (Taran, 1965)

- IS, and that [it] cannot NOT BE, (Hermann, 2004)

- that [it] is and that [it] is not not to be (Palmer, 2012)

4.2. NEGATION

The absurdum as a specific proposition

As I said above it is plausible to think about the absurdity as a certain proposition. This proposition is "truth is trivial"; or one may say, of course in a secondary reading, that "everything is true". The absurdity is saying that there is no difference between truth and the bare meaning, as if the difference between intention and fulfillment collapses. This would trivialize both acts of intending and experiences of truth. Thus one may say that in such a case everything, every intention, every supposition, every proposition can be said to be true; in any case no difference would be made. The difference, and the correlation, between subjectivity and the reality (wirklichkeit), between intention and truth is the more fundamental truth of the transcendental being; and to ignore it is *the* absurdity, namely the source of any other form of absurdity.

Absurdity, or *the* absurdum, must not be defined as the anti-truth. Falsity can be defined simply as non-truth. But absurdity in the strong sense of the term, so that can serve as a ground for the strong negation and for the strong sense of contradiction as well as for impossibility, needs more than this. Thus, the proper interpretation of the absurdum equates it with ignoring truth, or with trivializing truth, as described above. This interpretation differs from both equating the absurdum with the opposite of truth (this is just a combination of words without any significance at all) and equating it with the absence of truth (for absence of truth appears as of local significance and it is weaker than it be able to be a ground for the strong notions like absurdity or impossibility).

One may find also Gödel's philosophical view point in accordance with this attitude. As he said that the meaning of world consists in the separation between wishes and possibilities of fulfillment (Wang, 1996, p. 309), it is rational to suppose that to eliminate this difference is the meaninglessness, or is *the* absurdity.

Thus, in the case of the strong negation we can interpret $\neg p$ as $p \to \bot$, that is, if p is true then everything is true, or p would be true only if everything would be true.

So, the principle of *ex falso quodlibet* holds for the strong negation:

$$\bot \to p \qquad (4.6)$$

and

$$(q \wedge \neg q) \to p \qquad (4.7)$$

It says that if everything would be true then p also would be true. The formulas (4.5) and (4.6) are derivable from each other.[21]

Furthermore, the principle of excluded middle does not hold for strong negation, because it would say for any arbitrary p either p is true or it would be true only if everything would be true, and there is no evidence for such a claim.

Structural rule for the strong negation

To the structural rules of dialogical semantics for intuitionistic logic we add this structural rule:

> (SR-6) $\bot$ *is treated like an elementary proposition but if anyone uses it loses.*[22]

---

[21] However, I think that it is better to consider formula (4.5) as an axiom because it better represents the axiological content of the fundamental self-evident truth about absurdity.

[22] My primary suggestion was this : *Neither proponent nor opponent is allowed to assert* $\bot$. But after a discussion with Prof. Rahman I found his suggestion more expressive. The rationale behind both formulations are clear: that in the course of argumentation both parties are committed to the idea of truth and no one can trivialize the notion of truth. However the chosen formulation shows that the absurdum is really a proposition; it is not the case that $\bot$ is nothing and unsayable at all. Indeed one can assert it but it would mean that one gives up to strive toward truth or, in terms of the dialogical interaction, one loses.

Note: I recently noticed that Felscher (Felscher, 1986, p. 354) had briefly

4.2. NEGATION

The particle rules of dialogical semantics will remain same, except that there is no special rule for ¬.

From (SR-6) immediately proceeds that $\bot \to p$:

O	P	
	$\bot \to p$	(0)

There is no option for **O** to begin her attack. This formula may be added as an axiom which represents the rule (SR-6). But instead we may consider also the formula (4.5), which can be called Parmenidean axioma s the new axiom (added to the axioms of minimal, intuitionistic, logic). This is demonstrated as follows:

O		P	
		$(p \vee \bot) \to p$	(0)
(1@0)	$p \vee \bot$	$?_\vee$	(2@1)
(3①2)	p	p	(4①1)

Since **O** cannot assert $\bot$, she has to assert p in response to the attack 2, and then **P** can respond to the first attack and since there is no other option for **O** to do, **P** wins.

As we expected PEM is not valid:

O		P	
		$p \vee (p \to \bot)$	(0)
(1@0)	$?_\vee$	$p \to \bot$	(2①1)
(3@2)	p		

mentioned the idea formulated in my primary suggestion, that $\bot$ must not be asserted. I was not aware of his idea at the time of working on this part. However, I should acknowledge his primacy. Although my preferred formulation now is a little bit different one for the reason explained above.

In the move 2, **P** has to assert $p \to \bot$ because he cannot assert an atomic formula before that **O** does; and since he is obliged to respond the last attack he have to assert $\bot$ in response to the move 3 but in this case he loses, then **O** wins.

Also, elimination of double negation does not hold:

	O		P	
			$((p \to \bot) \to \bot) \to p$	(0)
(1@0)	$(p \to \bot) \to \bot$		$p \to \bot$	(2@1)
(3@2)	p			

According to the intuitionistic structural rules, just like the previous case, **P**, after the move 3, cannot use the atomic assertion p to respond the attack 1, because he has to respond the last attack, namely attack 3, and since he cannot respond this and also he cannot do any counterattack he loses.

Notice that our justification of ex falso here is different from the standard ones within the classical or constructive approaches. Van Atten's (van Atten, 2009) argument to show that ex falso can not be justified by means of proof interpretation is convincing. However, in contrast to him, I take this as a shortcoming of the so-called BHK interpretation. In general the fact that we admit intuitionistic logic does not mean that we are committed to the proof interpretation. According to the standard BHK interpretation ex falso holds because a function which is supposed to work there converting any proof for the absurdity to a proof for any given proposition does hold since there is no proof for the absurdity and such a function can not be shown that fails. But such an account just presupposes the validity of ex falso. We should be careful since the notion of ex falso may diffuse in the very notion of implication then to use such a notion of the latter in order to justify the former falls in circularity. The diffusion of ex falso into the implication is obvious in the classical, two valued, logic, for there an implication is simply *defined* as true

4.2. NEGATION

where the antecedent is false. But in intuitionism, and for us too as using the method of phenomenology, such a definition does not hold. So we should be careful not to use such an account when examining the vary case of ex falso. But it seems that such an account is at work in the BHK explanation of ex falso. It should be something like this: An implication is true when there is a function which converts any proof for the antecedent to a proof for the consequent, so an implication with absurdity as the antecedent is always true since absurdity has no proof and since we never have input we can assume an arbitrary function to satisfy our implication.[23]

To justify ex falso using the disjunctive syllogism is also not a proper way, since disjunctive syllogism depends, or at best is equivalent, to ex falso and if ex falso needs an explanation so does the disjunctive syllogism.

To examine alternative accounts of absurdity

I claimed that the treatment for ex falso introduced here is a novel one. In order to emphasize this and confront some possible objections I want to examine some well-known accounts of absurdity and show that the notion introduced here, i.e. the absurdity as meaning that there is no distinction between truth and mere meaning, is primitive in regard to other alternatives and that it serves to ground ex falso while others fail to do it.

1- There is a trend, mainly within the proof theoretical approaches toward logic, to consider absurdity as what there is no proof for it. And since, for them, the meaning of logical connectives are supposed to be given in terms of proof then the absurdum should be constituted as what lacks a proof (see for example (Troelstra and van Dalen, 1988). But such a definition is not a well-formed one at least from the phenomenological point of view because it confuses the intention with the fulfillment. To construct an intention (here a

---

[23]For further discussions of the intuitionistic account of ex falso see (van Atten, 2014).

propositional one) we can not refer to its fulfillment (here proof): an intention as such is independent of whether it is fulfilled or not so to define an intention as what can not be fulfilled gives just nothing. That is true that if we define absurdity like this then ex falso is groundless, in agreement with van Atten, but what indeed should be refuted is this very account.

2- Another option to define absurdity and then define negation on the basis of it is just to pick up a false proposition and consider it as $\bot$. It is also strange to claim ex falso on the bases of this—unless we have independently accepted ex falso then since that chosen proposition stands in conflict with a true proposition then under the supposition of that false proposition absurdity would hold and from this any arbitrary proposition follows. However, what is more strange is to define negation as leading to an arbitrary false proposition. It seems completely groundless to claim that for example it is not raining means that if it is raining then Socrates was not a philosopher. I mean it may be considered as possible to make such an inference on the basis of a negative judgment but to claim that this is the very meaning of that negative judgment is highly counter intuitive. Also if we take this then one may claim that for a positive proposition we have infinitely many negative ones with different meanings. Or, otherwise, for any proposition a specific false proposition should serve to be used as the consequent to formulate a negation of the former. All this suggestions (which have been introduced in some paraconsistent frameworks) are based on an extremely counter-intuitive account of negation, however they shows that indeed the original account is arbitrary and unacceptable. Moreover such a definition would be circular because if we want to use *a* false proposition exactly for being false then we should have been able to express its falsity then we should have negation before constituting the negation.

3- A suggestion is to consider absurdity as emptiness.[24] This is a

---

[24] In this paragraph I deal mainly with a thesis developed by Neil Tennent

4.2. NEGATION

good suggestion from those who want to avoid ex falso. Consider that we have a set of sentences Δ and a proposition φ, now we define $\neg\varphi$ as that the union of Δ and $\{\varphi\}$ has the empty set as succedent. In such an account what is taken as the peculiarity of absurdity is *blockedness*, then negation of a proposition is defined as making it explicit that to consider that proposition as true would block the reasoning. So just in the contrary to ex falso here nothing follows from absurdity. However ingenious such a suggestion is, it suffers from significant problems. First of all if we want to keep the rule of reflexivity namely $\varphi : \varphi$ then how a possible proposition would block the reasoning for it would have at least itself as a succedent? Then one should rule out monotonicity and impose some other restrictions to keep the original suggestion in a coherent way. Even if such a problem can be solved by some technical modifications, there is a still more deep philosophical problem. What is emptiness in regard to the proposition? We may assume that it is to say that there is no evidence at all, so the suggestion at work says that to negate a proposition means that if it would be true there would be no evidence to serve as ground for any judgment (that is, to add a given proposition to our reasoning would block any judgment). And it seems that from no evidence it can not be reached any arbitrary judgment so ex falso does not hold. But, to say that there is no evidence or, what is equivalent, there is no truth is itself a negative judgment and has been formulated on the basis of the notion of negation. Consider the proposition $\top$ as meaning that "there is a truth". Then by negating this we would have the proposition "there is no truth", so if we define $\bot$ as $\neg\top$ then to define the negation using $\bot$ would suffer from circularity.

Now again I emphasize that in the explanation of the absurdity that I just presented there is no flaw of the kind descried above.

(Tennant, 1999), that, to a large part, I agree with his philosophical considerations about the issue but his particular suggestion I found unsatisfactory as explained in the text.

First of all the absurdum is not an indexical or arbitrary proposition, it is really a specific proposition meaning that truth and meaning are the same or every intention (including every proposition) is fulfilled (or true). Notice that the meaning of the absurdum is not itself of a negative form so it does not presuppose the constitution of negation. And also it is a self-sufficient meaning and its explanation does not refer to the actual state of the realm of evidence or the act of fulfilling or proof constructing. That such a proposition should not be asserted is a metaphysical maxim. It is not defined by its lack of evidence or proof. That there is no proof for it is stated by an axiom (or a theorem), and more clearly by a structural rule, which itself has the meaning of this proposition as its constituent.

Some formal analyses of the new formulation of the rule for negation

Here we demonstrate also the principles of non-contradiction and double negation introduction.

	O		P	
			$(p \wedge (p \to \bot)) \to \bot$	(0)
(1@0)	$p \wedge (p \to \bot)$		$?_L$	(2@1)
(3ⓡ2)	p		$?_R$	(4@1)
(5ⓡ4)	$p \to \bot$		p	(6@5)

	O		P	
			$p \to ((p \to \bot) \to \bot)$	(0)
(1@0)	p		$(p \to \bot) \to \bot$	(2ⓡ1)
(3ⓡ2)	$p \to \bot$		p	(4@3)

4.2. NEGATION

Heyting's axioms. From (4.6), beside other axioms of minimal logic, the following formulas, namely Heyting's axioms for negation, are axiomatically derivable:

$$(p \to \bot) \to (p \to q)$$

$$((p \to q) \land (p \to (q \to \bot))) \to (p \to \bot)$$

Our modified structural rules grant them just as we might expect. Here we show the dialogue for the first one.

	O		P	
			$(p \to \bot) \to (p \to q)$	(0)
(1@0)	$p \to \bot$		$p \to q$	(2⒭1)
(3@2)	p		p	(4@1)

To show the core dialogue for the second formula, in which the proponent will win, is straightforward as well.

Discussion

Up to here we reach no new achievement but the intuitionistic logic. Then our strong negation is not a new one but the very intuitionistic negation.

If we consider minimal logic with the classical structural rule then add (SR-6) to it, we will obtain classical logic. Therefore, if the condition (SR-6) is to reveal the peculiarity of (strong) negation, then we see that there is no difference between intuitionistic negation and classical negation, rather the difference refers back to their structural rules. This point has been noticed before by (Rahman and Keiff, 2005). For instance, the structural rules of classical logic admit Peirce's axiom while those of intuitionistic do not, whereas there is no negation in that formula. If we explain this difference through the phenomenological viewpoint, the case is that both intuitionistic logic and classical logic begin with a same insight about negation. However, since intuitionistic logic function according to accurate rules it keeps the initial insight, but classical logic, because of unreliability of its rules violates the primary intuition. That is to say, our intuition of strong negation is that it means to imply the absurdity and absurdity is not admissible otherwise everything is admissible. Nevertheless, this intuition does not say that every proposition is either true or implies the absurdity. But classical logic, regarding its structural rule, reaches such a consequence, which is undesirable in respect to pure rules of reasoning.

4.2.5 Weak negation; to introduce a new rule

In order to have weak negation with aforementioned features we add a new rule. Let's show this new negation by the symbol $\sim$ and reserve the symbol $\neg$ for the strong negation. Now we introduce a new proposition and posit that $\sim p$ is equal to $p \to \pm$. The effect of the proposition $\pm$, we may call it *Withdrawal*, in the argumentation is represented by the following structural rule:

> (SR-7) *Withdrawal ($\pm$) is treated like an elementary proposition with a peculiarity:* **O** *may attack such an assertion if she has not asserted it before, and as a response* **P** *is allowed to respond to the previous unanswered attack or revise his answer to it, if possible.*[25]

---

[25] My primary formulation was as follows: " While one is attacked so that its response would be withdrawal, one may respond to the previous challenge." But after a discussion with prof. Rahman, the above mentioned formulation was preferred. It has an advantage that treats $\pm$ just like any other proposition which is capable to be asserted, however it according to its content may have an effect on the structure of the dialogue.

Notice that although I called this rule (SR-7), it cannot be employed together with (SR-5c). Indeed this latter would trivialize the former. So the suggested rule (SR-7) can be only considered as an extension for the intuitionistic logic, in

4.2. NEGATION

Of course as it is implied by the other structural rules one cannot repeat a response, namely in case relevant to (SR-7) one may move in response to the challenge which was not the last one, provided that that challenge has not been responded by the same move. This is a permission to revise but it works only if one is forced to assert withdrawal ($\pm$), and one may use it once and only in that occasion. The proposition $\pm$ is treated like other propositions except for the rule (SR-7), namely **O** may assert it and if **O** has asserted it **P** is also allowed to assert it.

Remember our example about the difference between the weak negation and the strong negation. There we said that in the case of the strong negation while one says not-p one claims to have evidence for the falsity of p as if one claims that p implies absurdity. Then if he is attacked by p he would have nothing to say (except to counterattack). But in the case of weak negation one only claims not to have evidence for p, so that if p would be the case a revision is needed, then if he is attacked by p (suppose that he has no option for counterattack) he is not forced to assert absurdity and thus loses, rather he is forced to revise. If he has nothing to revise (there is no open round or he is still not able to respond the last attack) it means that he is in a bad position and he has no choice but to lose.

If the *absurdum* says "absurdity is the case", roughly speaking, it can be said that *withdrawal* is a proposition which says "a revision is needed". The structural rule that is proposed is that if one is forced to assert this proposition he may instead do a revision.

For example, assume I assert "there is no angel" but I mean a weak negation, that is, I am not claiming that I have evidence for the non-existence of angels; rather, I am claiming I have no evidence in my epistemological system that angels exist. So "it is not that

---

contrast to the rule (SR-6) which in both classical logic and intuitionistic logic can substitute the rule for negation. However, we remind that the mention of classical logic is just for possible, technical interests; in the philosophical level our concern is intuitionistic logic—and, if possible and favorable, its conservative extensions.

there are angels" is equal to "if there are angels, then a revision is needed". (in contrast to "if there are angels, then absurdity is the case" which expresses a strong negation.) Now, if I confront the assertion saying there are angels, I may revise the last step of my reasoning.

Now according to the rule (SR-7), we see that PEM holds for the weak negation, as we intended:

	O		P	
			$p \vee (p \rightarrow \pm)$	(0)
(1@0)	$?_\vee$		$p \rightarrow \pm$	(2①1)
(3@2)	p		$\pm$	(4①1)
(5@4)	$?_\pm$		p	(6①1)

While being challenged for asserting $\pm$, **P** uses the assertion of p to respond the previous attack, move 1, and since no new move remains for **O**, **P** wins.

The following formula can be considered as the axiom which represents the semantic rule (SR-7):

$$p \vee (p \rightarrow \pm)$$

However, the following formula, elimination of double negation, is not proved as valid.

$$((p \rightarrow \pm) \rightarrow \pm) \rightarrow p$$

	O		P	
			$((p \rightarrow \pm) \rightarrow \pm) \rightarrow p$	(0)
(1@0)	$(p \rightarrow \pm) \rightarrow \pm$		$p \rightarrow \pm$	(2@1)
(3①2)	$\pm$			

4.2. NEGATION

In the move 4, instead of attacking, **O** responds the attack of **P**; and since no other move remains for **P**, he loses.

It would be clear that ex falso does not hold for the weak negation:

	O		P	
			$\pm \to p$	(0)
(1@0)	$\pm$			

Therefore, PEM holds and ex falso and elimination of double negation do not hold, as it was our primary intention. It is easy to show that non-contradiction and contraposition hold also for the weak negation.

4.2.6 A logic with two negations

We have now a logic with the strong negation and the weak negation embodied in a united system. The presented logical system can be considered as an expansion of intuitionistic logic. It contains all valid formulas of this latter and it rules out all formulas which are not generally valid according to it. In other words, it is not right to say that the logic just introduced contains intuitionistic logic as a fragment, as though the formulas not valid in this latter may be valid in the former. Rather, we can say that this new logic has intuitionistic logic as its core so that no disagreement, but just development, is allowed.

Any formula not containing $\pm$ is valid in this new logic iff it is valid in intuitionistic logic. For example the following formulas which are instances of the disagreement between intuitionistic and classical logic, but not concerning negation, are not valid too in the introduced logical system.

$$((p \to q) \to p) \to p$$

$$p \vee (p \to q)$$

In relation with classical logic, as for every formula ψ valid in the propositional classical logic, whether or not valid in intuitionistic logic, the formula $\neg\neg\psi$ is valid in the propositional intuitionistic logic, this is true also for the introduced logical system. It is sufficient to show that for the axiom that classical logic has in extra, namely $p \vee (p \to \bot)$, though itself not valid its double negation is valid according to our dialogical rules. See the dialogue on table 1.

We shall explain this phenomenon as follows. Since classical logic allows revision in every step, it can accept more claims as valid. But is it possible that a claim which is provable due to a revision be absurd at all, namely results in absurdity? The answer is no, for a claim which essentially entails the absurdity cannot be proven even with the revision in the course of argumentation. In the other words a claim valid in classical logic may be groundless (according to the philosophy which endorses intuitionism) but it would not be absurd (if the difference with the admitted reasoning would be just adding the possibility of revision). Therefore, if a claim is valid according to classical logic, we can, based on our more basic logic, say that at least the absurdity of that claim is absurd. That is, for every ψ valid in classical logic, $(\psi \to \bot) \to \bot$ is valid in intuitionistic logic and the logic introduced here.

However, it is not the case that if ψ is valid in classical logic, $(\psi \to \pm) \to \pm$ would be necessarily valid in our logic. In table 2 we show that the double weak negation of the axiom proper to classical logic is not demonstrable as valid according to the structural rule of the weak negation.

There **O** can answer to the attack 6 whereas **P** cannot use it to answer the attack 1 because he has to answer attack 5 first which he cannot, and since there is no counterattack possible, **P** loses.

We know that Brouwer, in order to show the unreliability of some rules of classical logic specifically $p \vee (p \to \bot)$, gave (weak and strong) counterexamples. These counterexamples are supposed to be

4.2. NEGATION

Table 1:

	O		P	
		$(((p \vee (p \to \bot)) \to \bot) \to \bot$		(0)
(1@0)	$(p \vee (p \to \bot)) \to \bot$		$p \vee (p \to \bot)$	(2@1)
(3@2)	$?_\vee$		$p \to \bot$	(4r3)
(5@4)	p		$p \vee (p \to \bot)$	(6@1)
(7@6)	$?_\vee$		p	(8r7)

Table 2:

	O		P	
		$(((p \vee (p \to \bot)) \to \mp) \to \mp$		(0)
(1@0)	$(p \vee (p \to \bot)) \to \mp$		$p \vee (p \to \bot)$	(2@1)
(3@2)	$?_\vee$		$p \to \bot$	(4r3)
(5@4)	p		$p \vee (p \to \bot)$	(6@1)
(7r6)	$\mp$			

evidence for that the mentioned rules are without adequate evidence. Now if for every ψ valid in classical logic, $(\psi \to \pm) \to \pm$ were valid, it was implying that there is no evidence for the evidence-less-ness of ψ, while the counterexamples show that in fact there is.

Certain important formulas

Now, we continue to investigate the behavior of this new system in respect with some of significant logical issues.

Double Negation. As we said before, elimination of double negation does not hold, but it happens that the following is valid:

$$((p \to \pm) \to \bot) \to p$$

	O		P	
			$((p \to \pm) \to \bot) \to p$	(0)
(1@0)	$(p \to \pm) \to \bot$		$p \to \pm$	(2@1)
(3@2)	p		$\pm$	(4①1)
(5@4)	$?_\pm$		p	(6①1)

Since in the move 6, **P** has to revise a previous move, or respond a previous attack, he uses p which is now asserted by **O** to respond the challenge 1. Then no option to move remains for **O** and she loses.

Explanation. This is interesting, but not very surprising. Remember that we said that elimination of double negation is a result of a confusion between two kinds of negation. So, we might expect that a combination of weak and strong negation may satisfy something like the mentioned principle. The above formula, namely $\neg \sim p \to p$, is not an axiom and it can be derived from our axioms.

4.2. NEGATION

It can be informally explained as follows. According to the Parmenidean axiom, if in a disjunction one side entails the absurdity that is the other which is the case. So, as we have the disjunction $p \vee (p \to \pm)$, then if the left side implies the absurdity, the right side is the case[26], that is: $((p \to \pm) \to \bot) \to p$.

This theorem says that if having no evidence for p leads to absurdity then p is true; or if it is absurd for p to be subject of a revision then it is the case. While, again, $\neg\neg p \to p$ says if it is absurd that p leads to absurdity then p is the case, which is not generally true. And $\sim\sim p \to p$ says if that p forces a revision forces a revision then p is the case (or that there is no evidence that there is no evidence for p implies that p is the case), which is also unjustified.

Elimination of double negation is only valid for the above mentioned order. $\sim \neg p \to p$ is not valid. It is simple to demonstrate it semantically. Clearly, if there would be no evidence that a certain proposition leads to absurdity it would not necessarily imply that that proposition is true. For example we have no evidence that Goldbach's conjecture leads to absurdity; this does not serve a ground to take the conjecture as true. The mentioned formula is valid in the other way around, namely if p is true then there would be no evidence for that it is absurd. This form of double negation introduction ($p \to\sim \neg p$) is valid beside the others we have already seen($p \to \neg\neg p$ and $p \to\sim\sim p$). But this is not valid: $p \to \neg \sim p$. Let's examine it.

---

[26] Also if the right side implies the absurdity, the left side is the case. This is evident also by a more basic way, that is the strong negation entails the weak negation. See below.

O		P	
		$p \to ((p \to \pm) \to \bot)$	(0)
(1@0)	p	$(p \to \pm) \to \bot$	(2Ⓡ1)
(3@2)	$p \to \pm$	p	(4@3)
(5Ⓡ4)	$\pm$	———	

Thus, it is not generally true that if p then having no evidence for p would result in absurdity. If p is true of course it is wrong that there is no evidence for p (or p is subject to revision), but this is not strongly wrong, not that it would result in absurdity. In the other words, if p is the case, then only that it would be absurd is absurd, and that we have no evidence for it (i.e. we do not know that it is true) would not be absurd. This point is of interesting philosophical and epistemological implications, which are not of course the subject matter of the present investigation.

Since there are two negations, four cases of double negation are possible, namely $\neg\neg$, $\sim\sim$, $\sim\neg$ and $\neg\sim$, among which for the first three introduction holds but elimination does not hold, while for the last one elimination holds but, as we just saw, introduction does not hold. Therefore, for none of the four cases of double negation, both elimination and introduction hold.

Strong implies weak. In the relation between the weak negation and the strong negation, we may also except that the latter implies the former. This can be shown as valid according to our rules:

O		P	
		$(p \to \bot) \to (p \to \pm)$	(0)
(1@0)	$p \to \bot$	$p \to \pm$	(2Ⓡ1)
(3@2)	p	p	(4@1)
	———		

4.2. NEGATION

In fact, it can be seen that that the above formula is easily derivable because $\bot \to \pm$, while this latter is a special case of *ex falso*.

De Morgan laws. We know that from their four cases three hold in intuitionistic negation, and thus for our strong negation, and one does not. This is the formula which is not generally valid:

$$((p \land q) \to \bot)) \to ((p \to \bot) \lor (q \to \bot))$$

All of the four cases of the De Morgan's laws are valid for weak negation. In the following we show the invalidity of the above formula in order to compare it with the case of the weak negation. See table 3.

But, using the rule (SR-7) we will see that the following form is valid for weak negation, as shown in table 4.

$$((p \land q) \to \pm)) \to ((p \to \pm) \lor (q \to \pm))$$

This can be explained in this way. While p and q both together lead to absurdity, there is no implication that one of them by itself would lead to absurdity. However, if there is no evidence that p and q both are the case, then it is clear that at least one of them lacks evidence.

Negations and implication In the classical logic $p \to q$ is equivalent to $q \lor \neg p$, then negation and implication together can define other connectives. On the other hand implication is reducible to negation plus one of disjunction or conjunction. But in the intuitionistic logic if we have $q \lor \neg p$ we would have $p \to q$, but not the other way around. Now it is worthwhile to examine that what is the relation between the weak negation and implication.

Is implication would be reducible to disjunction plus weak negation or weak negation beside strong negation? Interestingly the answer is no.

Table 3:

O		P	
	$(p \wedge q) \to \bot$	$((p \wedge q) \to \bot) \to ((p \to \bot) \vee (q \to \bot))$	(0)
(1@0)	?$_\vee$	$(p \to \bot) \vee (q \to \bot)$	(2①1)
(3@2)	?$_\vee$	$p \to \bot$	(4①3)
(5@4)	p	$p \wedge q$	(6@1)
(7@6)	?$_R$		

Table 4:

O		P	
	$(p \wedge q) \to \pm$	$((p \wedge q) \to \pm) \to ((p \to \pm) \vee (q \to \pm))$	(0)
(1@0)	?$_\vee$	$(p \to \pm) \vee (q \to \pm)$	(2①1)
(3@2)	?$_\vee$	$p \to \pm$	(4①3)
(5@4)	p	$\pm$	(6①5)
(7@6)	?$_\pm$	$q \to \pm$	(8①3)
(9@8)	q	$\pm$	(10①9)
(11@10)	?$_\pm$	$p \wedge q$	(12@1)
(13@12)	?$_L$	p	(14①13)
(15@12)	?$_R$	q	(16①15)

4.2. NEGATION

In fact instead of the classical theorem $(p \to q) \equiv (q \vee \neg p)$ we have:

$$(q \vee \neg p) \to (p \to q)$$

$$(p \to q) \to (q \vee \sim p)$$

while the following are not valid:

$$(q \vee \sim p) \to (p \to q)$$

$$(p \to q) \to (q \vee \neg p)$$

We will demonstrate the cases for the weak negation:

	O		P	
			$(p \to q) \to (q \vee (p \to \pm))$	(0)
(1ⓐ0)	$p \to q$		$q \vee (p \to \pm)$	(2ⓡ1)
(3ⓐ2)	$?_\vee$		$p \to \pm$	(4ⓡ3)
(5ⓐ4)	p		p	(6ⓐ1)
(7ⓡ6)	q		$\pm$	(8ⓡ5)
(9ⓡ8)	$?_\pm$		q	(10ⓡ3)

In the move 8, **P** responds to the last attack (move 5) and asserts the withdrawal. So when his move is challenges he is able to respond a previous attack, and he responds the move 1 which now he is able to respond. Since there is no more move possible for the opponent, **P** wins.

We know that either p is the case or there is no evidence for p, now if we admit that p implies q it is easily seen that either q is the case or there is no evidence for p.

O		P	
		$(q \vee (p \to \pm)) \to (p \to q)$	(0)
(1@0)	$q \vee (p \to \pm)$	$p \to q$	(2⊙1)
(3@2)	p	$?_\vee$	(4@1)
(5⊙4)	$p \to \pm$	p	(6@5)
(7⊙6)	$\pm$	———	

There is no other possibility for **P** and he loses. From ex falso it follows that if $\neg p$ then $p \to q$ would hold, but it is not true that if we have no evidence for p then we would have evidence that p would imply q.

Peirce's law The formula called Peirce's axiom, or as Russell called it[27], the principle of *reduction*, is not, as discussed before, valid in our system. However, similarly we do have $((p \to \pm) \to p)) \to p$, that is we have $(\sim p \to p) \to p$ while we do not have $(\neg p \to p) \to p$. If $\neg p$ itself lead to p it shows that $\neg p$ is not the case but as PEM does not hold for the strong negation we cannot infer that p is true. Whereas, for the weak negation we know that either p or $\sim p$ is true then if we have $\sim p \to p$, then in any case p would be true.

4.3 Necessity and Strict Implication

The objective of the present section is to introduce a dialogical explanation of *necessity*. Nowadays it seems to be very difficult to think about the modalities necessity and possibility outside of the possible worlds framework. But the question is whether such a framework is really to reveal the meaning of these modalities or it is just a technical method to serve as a semantics? If the latter is the case we should be careful not to consider what follows from our

---

[27](Russell, 1903). See, for a discussion on the relation of this formula with other classical theorem specially PEM, (Mints, 2006)

4.3. NECESSITY AND STRICT IMPLICATION

framework as the metaphysical character of the notions in question. The fact that the possible world semantics (PWS hereafter) as a logical tool is very powerful is doubtless. However in order it to be taken as sufficient for the meaning explanation, something more than being technically fruitfulness is required. Of course there are philosophical arguments for or against the genuineness of PWS and its proper formulation. One alternative is the inferentialist approach to modality in which the meaning of, say, necessity is supposed to be given through an introduction rule.[28] However, this approach has its own philosophical problems, more importantly it somehow equates necessity with the logical necessity namely it introduces necessity only for those formulas which are valid according to some logical rules, thus neglects any primitive significance for necessary judgments.[29] My aim is to introduce another method to serve as semantics for necessity which is able to properly represent the intention behind this modality.

I just want to outline a new approach which, once sufficiently developed, would provide us with a tool to deal with necessity and the issues around them. In this respect I invite the reader not to think in terms of possible world semantics about necessity and possibility while studying this note. If you keep in mind the PWS account of necessity and expect some new technical achievements, there is nothing remarkable in this note. However, The aim is to demonstrate that there is at least an alternative way to understand modalities; and there is no need to assume PWS as intrinsic to the concept of necessity and possibility.

---

[28]Such an approach is developed for example by Stephen Read (Read, 2010) (Read, 2008).

[29]Beside the mentioned criticism, inferentialism uses a kind of "labelled" deductive system in order to develop a semantics for modality while such a system does reflect a kind of possible world attitude. But even putting such an objection aside, one may still has doubts about the appropriateness of the inferentialist theory of meaning, doubts of the kind that have been explained by Arthur Prior.

A dialogical semantics for modal logic has been developed by Rahman and Rückert (Rahman and Rückert, 1999). However, the main idea of that work is rather to adapt the PWS into the dialogical framework, so that, though very fruitful in certain respects, it rather follows the PWS interpretation of modalities instead of offering a proper explanation. In the present investigation, I am going to use the framework of dialogical logic to develop a modal logic and to explain the nature of necessity, beginning with the Leibnizian account of this notion—which will be explained below.

I begin with a brief discussion to explain the Leibnizian thesis. Then I will formulate new dialogical rules, on the basis of the Leibnizian thesis, to represent necessity. I will sketch out the main characteristics of the logic so obtained. It will be similar to **S4** with both classical and intuitionistic versions. At the end I discuss the philosophical implications of the idea so explained. I will argue for the appropriateness of the introduced interpretation of necessity in respect with the intention behind this modality as it is recalled in the course of reasoning and argumentation. I will also mention some possible, further works to use this idea in order to develop a comprehensive dialogical framework to deal with modalities and the issues around them.

4.3.1 Thesis: the Leibnizian account of necessity

Necessity is in general conceived of in contrast to possibility. Then it is indispensable to take a, though cursory, look to this latter. It is customary to take possibility and necessity as interdefinable. But for the moment let us put this assumption aside and see if there is any independent (though related) notion of each one.

Now, again let us suspend any metaphysical sense of "possibility" and see what is the intention(s) of using such a modality in a discourse or, in other words, what is supposed to be observed

4.3. NECESSITY AND STRICT IMPLICATION

when we modify a judgment as possible. It seems that there is an ambiguity around this term and there are at least three notions of "possibility":

1. Plausibility, Intelligibility

2. Contingency, namely when we speak about a fact but indicate that its realization is not necessary

3. As that is not necessarily ruled out, whether the subject matter is the case or not

Possibility in its turn is conceived in contrast to either absurdity or reality. If we conceive absurdity as violating the basic rules of rationality then possibility means what is conceivable or intelligible in some way. On the other hand in some occasions we use the word possibility to indicate that we are not speaking strictly about reality rather we speak about an objectivity regardless to its realization. The notion of contingency is also related here because we need it to indicate that a fact could not take a place or it's absence is intelligible. As it can be seen the three notions are interrelated but they are not same. The question that which one of them is of priority though very important is beyond the task of this investigation. On the other hand we may argue that, in contrast to "possibility", there is a single notion signified by the term "necessity".

In the present work I begin with a Leibnizian thesis as the point of departure. I am not going to argue for the aptness of this thesis. I just want to develop a logical formulation of necessity based on such a thesis and afterwards discuss its possible philosophical advantages over other attitudes.

Leibniz in several occasions defines necessity in contrast to contingency in a way that can be formulated as the following:

> Necessary truths are those that are derivable, through finite steps, from the fundamental truths; while the steps

of derivation for the contingent truths are infinite.[30]

It is quite common to understand necessity as unconditional truth, namely that kind of truth which is unchallengeable according to its nature; and to deny it is in some sense is equal to give up the rational attitude. Now Leibniz says that anything derivable from such truths is also necessary. For Leibniz every fact is ultimately based on the primary truths. However, those facts that we consider as contingent are involved in a continuity of co-existing phenomena so that the chain of reasons behind them are infinite; and from the point of view of human being to trace back the reasons to the fundamental truths can not be done.

Here we have two key notions, one is the set of fundamental truths and the other is the notion of derivation. To describe this in a logical way we can say that we begin with a set of fundamental, or canonical, truths and we have a consequence relation that can be seen, in a close look, that is different from the ordinary implication connective which is formulated in the classical or intuitionistic logic. Indeed the important point is this very difference and that we have to try to introduce a new implication to satisfy the notion of derivation recalled in the thesis. But what is the difference?

In both classical and intuitionistic logic we have the following axiom:

$$p \to (q \to p) \qquad (4.8)$$

It says that if some proposition is true it is implied by any arbitrary proposition. Now if we take that notion of derivablity which is used in the aforementioned Leibnizian thesis as represented by such a notion of implication used in the above axiom, every truth would turn out to be necessary, for a true proposition would be implied by any proposition including the canonical truths so it can be considered as derived from them and should be considered as necessary. Then we need another conditional than what validates 4.8.

---

[30]See *Monadology* paragraphs 33–36, (Leibniz, 1989, p. 646).

4.3. NECESSITY AND STRICT IMPLICATION

One may object here that the validity of 4.8 is not due to the character of the material implication but to a structural rule; and then what should be modified in order the Leibinzian thesis not to fall in triviality is not the kind of conditional. Such an argument goes as follows:

The validity of 4.8 relies on two things, namely the semantics of $\to$ and weakening.

$p \vdash p$ (axiom)

$p, q \vdash p$ (weakening)

$p \vdash q \to p$ ($\to$ intro)

$\vdash p \to (q \to p)$ ($\to$ intro)

So, the mentioned argument claims, that perhaps what should be taken as problematic is the weakening not the semantic of the material implication; for there are other proofs in which the same semantic for the implication is used and it seems not be problematic for our main thesis. For example:

$(p \land q) \vdash (p \land q)$ (axiom)

$(p \land q) \vdash p$ ($\land$ elimination)

$\vdash (p \land q) \to p$ ($\to$ intro)

Given that $(p \land q) \to p$ is not problematic, this does not mean that the issue does not come from the semantic of the material implication. In a close look we will see that the application of the introduction rule in the two proofs are different. In the former just one of the premises is chosen as the antecedent of the introduced implication, but in the latter all premises (which is a single formula here) are brought as the antecedent. The difference is exactly here. I argue that for the Leibnizian account of necessity the former case would be problematic if we take the implication as representing the relation of derivablity. There is no reason to think that weakening is undesirable and indeed it is not—as our discussion will show. But the problem is that the semantic of material implication is not suitable in order to formulate our idea about necessity. Thus we need another conditional. This does not mean that the material implication is unsuitable at all or that it should be replaced. Rather I am arguing that we can expand our logic by adding a new connective so that necessity can be defined on the basis of this new connective.

The formula 4.8 is sometimes called *the positive paradox of material implication*. Here, I am not claiming that it is "paradoxical". Instead I argue that the material implication, due to the fact that it validates 4.8, is not capable to be used in formalizing the above mentioned idea in order to represent a logical account of necessity. Here we need another kind of implication. It could be a *relevance implication*, since one of the motivation of relevance logic is to avoid 4.8. Also it could be a *strict implication* or some other possible conditionals. However, I do not aim just to pick out one of the conditionals which have been introduced in the literature; rather I intend to follow the main idea to formulate a conditional capable to satisfy our motivation, and, as to be explained in the next section, it turns out to be the very *strict implication*.

The relation between the strict implication and the modalities is already known in the literature. However, it is customary to take necessity or possibility as primitive then define the strict implication by means of them.[31] Accordingly it is usual to give a possible world semantics for the strict implication. My approach here is the other way around. I think, in accordance with the Leibnizian thesis, the notion of strict implication has priority over, at least, necessity. Moreover, I argue that the dialogical approach could better clarify the nature of the strict implication and serves as a semantics for it in a way which is of less metaphysical presuppositions than the other alternative approaches.

In the following I describe what is intended from the strict implication and what its logical features should be. Then I will formulate it in a logical way using the dialogical semantics.

---

[31] Although this was not the primary intention of C. I. Lewis who for the first time formulated the notion of strict implication.

4.3.2 Strict implication

The key point in our required relation is that when we say φ is strictly derivable from ψ, we mean that in order to derive φ we are allowed, if needed, only to use the claims stated by ψ; we are not allowed to use other hypothesis or to look at the truths given outside of the assumption; however there is no obligation to use the assumption, namely if ψ itself is an unconditional truth then it has to be present everywhere and it is legitimate to say that ψ is strictly implied by φ. In the other words, in our strict implication the consequence in order to be true does not need something more than the antecedent.

In order to stress the difference between the material implication and the strict one, we may analyze the following formulas:

$$(\psi_1 \to (\psi_2 \to \varphi)) \to ((\psi_1 \wedge \psi_2) \to \varphi) \tag{4.9}$$

$$((\psi_1 \wedge \psi_2) \to \varphi) \to (\psi_1 \to (\psi_2 \to \varphi)) \tag{4.10}$$

Let us read the formula 4.9 as saying that if it is true that if ψ_1 holds then ψ_2 would imply φ, then if both ψ_1 and ψ_2 hold then φ would also hold. This is quite intuitive. Now let us see the conditional from the other side, namely the formula 4.10, assume that if both ψ_1 and ψ_2 hold then φ would also hold, now would it imply that if ψ_1 is true then if ψ_2 holds then φ would also hold? This also seems to be true. However if we use the notion of assumption, instead of premises, the case for the formula 4.10 would be different: it says if from the assumption ψ_1 and ψ_2 it is implied that φ holds, then from the assumption ψ_1 it is implied that φ is derived from ψ_2. If we accept this latter reading and take formula 4.10 as valid then we are using the material implication. However if by implication we understand a strict one, formula 4.10 would not be in general valid, for from the fact that A is strictly derivable from *both* B and C, it can not be concluded that if B holds then A would be strictly derivable from *only* C. So both the material implication and the strict implication validate the formula 4.9, but the formula 4.10 is admitted for the general implication and not for the strict implication.

Assume that the formula 4.10 is valid and substitute ψ_1 by p, ψ_2 by q and φ again by p, so we will have:

$$((p \wedge q) \to p) \to (p \to (q \to p)) \tag{4.11}$$

The antecedent, namely $(p \wedge q) \to p$ is evident, so as a consequence we have:

$$p \to (q \to p) \tag{4.12}$$

It means that the above formula is valid for the material implication; and it is usually considered as one of the paradoxes of material implication.[32] Indeed it is one of the Hilbert's axioms for the classical logic and also one of the Heyting's axioms for the intuitionistic logic. Then as a criterion we can say that after formulating a strict implication according to our motivation, it should not validate the above formula. It is important to notice that in order to introduce a strict implication there is no need to revise the whole logical system. We can have a conservative extension of either classical logic or intuitionistic one. This will be clear after explaining our semantic rules.

Now, since we are clearer about our notion of strict implication, it would be good to make certain remarks in this respect before going to the formal investigation.

1- According to the mentioned Leibnizian idea every truth that is implied by some fundamental truths is considered as necessary.[33]

---

[32] Though there is no paradox by itself here. Paradox would arise if one takes the material implication as the strict one, namely confuses the two and expect that it satisfies the peculiarities of the strict implication.

[33] *Quand une vérité est nécessaire, on en peut trouver la raison par l'analyse, la résolvant en idées et en vérités plus simples, jusqu'à ce qu'on vienne aux primitives.* (*Monadology, 33*)

4.3. NECESSITY AND STRICT IMPLICATION

That is if q is a fundamental truth, p would be necessary provided that:

$$q \rightarrow_* p$$

As we discussed above, the connective $\rightarrow_*$ cannot be replaced with a material implication ($\rightarrow$) for it does not satisfy our primary idea. We are going to introduce a rule in order to represent that notion of implication which is used in our definition of necessity. Moreover, the connective $\rightarrow_*$ is not a relevance implication for here we do not care about the relevance between antecedent and consequent. All we need is that in order to derive p we do not use something other than q (the idea is that we do not refer to the facts but to some fundamental truths); and it is possible that there would be no need to q itself so that p can be irrelevant to q.[34] The implication that we use in lieu of $\rightarrow_*$ is the strict implication as we explained above and we are going to formulate it by means of a semantic rule.

2- In some occasions Leibniz gives a seemingly different definition for the necessity. However, here I want to show that this second definition is also based on the strict implication, and if you prefer the classical logic the two definitions would turn out to be the same. Since I prefer intuitionistic logic I have chosen the above one. The second definition can be stated as follows:

> It is necessary what its contrary implies a contradiction.[35]

---

[34]In most parts the logical system we are going to introduce is similar to a relevance logic, specially to logic **E**, however the original motivation is different and there would be some significant differences in valid formulas.

[35]Leibniz gives such a definition in the paragraph 13 of *Discourse on Metaphysics*:

> In order to meet the objection completely, I say that the connection or consecution is of two kinds; one is absolutely necessary, whose contrary implies contradiction, and that deduction occurs in the

That is, p is necessary if

$$\neg p \rightarrow_* \bot^{36}$$

But it should be obvious that what Leibniz means here by "implies" is the strict implication not the material one, for otherwise every truth would be necessary. Assume that p is true, so in accordance with an axiom of classical and intuitionistic logic, we will have:

$$\frac{p}{\neg p \rightarrow p}$$

---

eternal truths like the those of geometry; the other is necessary only ex hypothesi, and so to speak by accident, but in itself it is contingent since the contrary is not implied. This latter sequence is not founded upon ideas wholly pure and upon the pure understanding of God, but upon his free decrees and upon the processes of the universe.

(Leibniz, 1989, p. 310)

It should be said that this definition is in essence the same as stated before. However, since we want to do an analysis in a rigorous way the formal difference may be of an importance; and as I say in the text I prefer the former formulation.

[36]This may seem to be similar to the definition of necessity that Williamson (Williamson, 2010, p. 83) suggests on the basis of the notion of counterfactual conditional introduced by Lewis (Lewis, 1973):

$$\Box p \equiv (\neg p \Box\!\!\rightarrow \bot)$$

However, there is no real connection here. First of all, because counterfactual conditional is not a strict implication as we are employing here. Lewis himself shows this and say:

We shall see that the counterfactual cannot be any strict conditional. (Lewis, 1973, p. 4)

Moreover, counterfactual is of a temporal aspect which is irrelevant to our thesis. More importantly, counterfactuality as introduced by Lewis is based on the possible world framework which we deliberately avoid. Then, although the two formula may seem similar, the meanings of them are essentially different.

4.3. NECESSITY AND STRICT IMPLICATION

From the other hand, by means of self-implication we have:

$$\frac{}{\neg p \to \neg p}$$

Putting these two together, it follows that:

$$\frac{p}{\neg p \to (p \land \neg p)}$$

It means that if p is true then its negation would lead in a contradiction. So if the implication in the above definition is understood as the material implication, every truth should be considered as necessary. This shows that we really need a special kind of implication which blocks such a triviality. It should be clear that Leibniz's intention is that kind of implication which does not use other than the assumption declared. For example, if from the assumption that the sky is not blue, and not using any other information, a contradiction is derived then that the sky is blue would be necessary. So, both definitions, though they may seem different, are based on a same and more fundamental notion which is the strict implication.[37]

3- When we use the expression "necessary and sufficient condition" we must have a concept of a strict implication or a strict conditional in mind, otherwise such a term would be pointless. In general when we assert that p implies q it means that p is conditioned by q, so in order p to hold q must hold. And when we assert s implies p it means that s is sufficient for p to hold though it is possible that p holds and s does not. Now if we stop here and do not go into the analysis of the implication at work, namely if we take the material implication as the admitted one then as conclusions we would have that every truth is a necessary condition for any arbitrary proposition, any

---

[37] However, this latter definition is primarily to define $\Box \neg \neg p$. Since in the classical logic we have $\neg \neg p \leftrightarrow p$ this would be equal to $\Box p$. But in the intuitionistic logic we only have $p \to \neg \neg p$, so there is a difference between the two definitions and the one I have chosen is the appropriate one if we want to formulate necessity in a positive way.

arbitrary proposition is a sufficient condition for any truth and any arbitrary proposition is a necessary condition for a false proposition. But these are not so useful principles. Indeed we intend to recall the strict conditions when we use the term "necessary and sufficient condition", namely while saying s is sufficient for p we mean that from s itself, without referring to any other contingent fact, p is derived as a conclusion and while saying q is a necessary condition for p, we mean from p itself and without referring to any contingent fact it follows that q holds.

4- It is worth mentioning that neither this new connective nor the material implication is to represent the merely logical consequence relation. That is if I interpret $p \dashv 3\ q$ as q is strictly implied by p or q is derivable from p this does not necessarily mean that q is *logically* derivable from p. In other words $\dashv 3$ is not to represent $\vdash$. Although if $p \vdash q$ we would also have, as will be shown in the rest of the present work, that $p \dashv 3\ q$, it is not necessarily the case for the other way around. Namely we are not dealing with necessity as a metalogical notion, rather as a universal kind of judgment modification which can be treated within a logic. The kind of implication, that we should searching for in order to indicate the expression that there is a way for reasoning from the antecedent to the consequent, is not the so-called object language counterpart of the inferential or logical consequence relation.[38]

A new rule for the strict implication

In the following I use the notation $\dashv 3$ to indicate the strict implication. As we saw, the dialogical rule for the material implication is based on the idea that if one asserts an implication then one should be able to assert the consequent if the challenger asserts the antecedent. Now I follow this idea and I add a restriction that once a strict conditional

---

[38]That is why our approach is different from any proof-theoretical account of necessity, as for instance developed by Stephen Read (Read, 2010), although they too decline the possible world framework.

4.3. NECESSITY AND STRICT IMPLICATION

is challenged by asserting the antecedent (or the assumption), the defender is not allowed to refer what have been asserted before this challenge, but he/she is allowed to use what have been accepted to be strictly implied by any thing that he is allowed to assert under the assumption. The rule corresponding to this observation goes as follows:

> *A formula in the form $\psi \dashv \phi$, i.e. strict implication, can be attacked by asserting ψ and the response is to assert ϕ; but there is an extra condition:*
>
> *If in the move n, **X** asserts a strict implication and **Y** attacks it in the move m, m > n, then in the following moves **X** could not refer to any move l, l < m, of **Y** except that it was a strict implication; if **X** is the proponent this also means that he is not allowed to use an elementary assertion which is made by the opponent only before the move m.*[39]

We said that the defender of a strict implication can not refer to any move before the attack unless it was itself a strict implication. This grants that if **X** is able to assert α and if **Y** has accepted that β is strictly implied by α, **X** would be able also to request β and use it.

Now we can see the invalidity of the formulas 4.8 for the strict implication:

---

[39]We know that in general the proponent can not assert an atomic unless it has been asserted by the opponent before. Now if the opponent attacks a strict implication asserted by the proponent, the proponent is obliged to *forget* all atomics asserted before this attack.

	O		P	
			$p \dashv (q \dashv p)$	(0)
(1@0)	p		$q \dashv p$	(2⊤1)
(3@2)	q		———	

P has no other move because he can not use the elementary assertion p since it is asserted before the move 3, and also no counterattack is possible, then he loses.

Another formula which is problematic in respect with the strict implication, $\neg p \rightarrow (p \rightarrow q)$, can be demonstrated, as it ought to be, that is not valid according to our rule:

	O		P	
			$\neg p \dashv (p \dashv q)$	(0)
(1@0)	$\neg p$		$p \dashv q$	(2⊤1)
(3@2)	p		———	

P loses because he can not use the elementary assertion p in order to attack the move $\neg p$ since it is asserted before the move 3; thus **P** has no winning strategy and the formula is not valid.

However a self-implication in the form $(p \dashv q) \dashv (p \dashv q)$ is valid as it should be:

	O		P	
			$(p \dashv q) \dashv (p \dashv q)$	(0)
(1@0)	$p \dashv q$		$p \dashv q$	(2⊤1)
(3@2)	p		p	(4@1)
(5⊤4)	q		q	(6⊤3)
	———			

Here in contrast to the previous dialogue, **P** can use the elementary assertion p in order to attack the move 1 because, although it is

4.3. NECESSITY AND STRICT IMPLICATION

asserted before the move 3, it is a strict assertion; thus **P** wins and the formula is valid.

Although $\neg p \to (p \strictif q)$ is not valid, *ex falso* in the form $(\neg p \wedge p) \strictif q$ is valid for strict implication. This latter sometimes is considered as one of the paradoxes of the strict implication. However, there is nothing unintuitive here. Any argument in favor of *ex falso* for material implication should also work for the strict implication, for such an argument would progress in a way to show that from the absurdity itself, or from the contradiction itself, any arbitrary proposition follows; it means that the connection here is strict. For the strict implication it is not true that from any arbitrary false proposition every thing follows, but the original form of *ex falso* which is to say if absurdity holds then any arbitrary proposition would be true is valid. There are of course some doubts about the validity of *ex falso* in general. I have argued for the correctness of this principle in sec. 4.2.4. My point here is that if *ex falso* is to be considered valid for the material implication, then it is also valid for the strict implication. Remember that the intention behind the strict implication was to reject the formula 4.8, therefore although we accept the following

$$(\neg p \wedge p) \strictif q$$

it does not lead us to accept the validity of the followings:

$$\neg p \to (p \strictif q)$$

$$p \to (\neg p \strictif q)$$

That one of these latters is not valid has been shown above. The other goes in a similar way. The validity of *ex falso* itself, $\bot \strictif q$, is easy to demonstrate.

Another phenomenon which is sometimes called as a paradox of strict implication is that if ψ is a valid formula then for any arbitrary p, $p \strictif \psi$ is also valid. However, there is nothing paradoxical here and this phenomenon does completely fit our intention behind the strict implication. Indeed our semantics must grant this phenomenon and it does so. Notice that by strict implication we intend that the consequent does not need more than the assumption to be true. Now if ψ is valid, namely it is a formal truth, then in the formula $p \dashv 3\ \psi$, it does not need anything other than p to be true, neither does it need p itself. It means that if we accept that something is true only according to the rules of pure reasoning, it should be true under any assumption.

Here I show the dialogue for the case of $p \dashv 3\ p$. This is valid in general, so $q \dashv 3\ (p \dashv 3\ p)$ must be valid:

	O		**P**	
			$q \dashv 3\ (p \dashv 3\ p)$	(0)
(1@0)	q		$p \dashv 3\ p$	(2①1)
(3@2)	p		p	(4@1)

Here, in the move 2 **O** attacks a formula with strict implication, and in response to it **P** has not refer to any move before the move 2. He uses the atomic asserted by **O** in the move 2 itself in order to respond. Since there is no possibility for **O** to attack, **P** wins and the thesis is demonstrated to be valid.

It is worth mentioning that the strict implication is transitive, namely the following formula is valid:

$$(p \dashv 3\ q) \dashv 3\ ((q \dashv 3\ r) \dashv 3\ (p \dashv 3\ r))$$

This is demonstrated in table 5. There in the moves 6 and 8, **P** is able to attack, respectively the moves 1 and 3 because they are strict implications.

Now, having explained the logical character of the strict implication, I shall go to investigate the logical character of the canonical truths in order to formulate necessity. Before that I should emphasize that the material implication and the strict implication can both

4.3. NECESSITY AND STRICT IMPLICATION

O		P	
			$(p \strictif q) \strictif ((q \strictif r) \strictif (p \strictif r))$	(0)
(1@0)	$p \strictif q$	$(q \strictif r) \strictif (p \strictif r)$	(2⊩1)
(3@2)	$q \strictif r$	$p \strictif r$	(4⊩3)
(5@4)	p	p	(6@1)
(7⊩6)	q	q	(8@3)
(9⊩8)	r	r	(10⊩5)

Table 5:

be used in a same logical system and they are not reducible to each other; from the strict implication, material implication follows and not vice versa.

4.3.3 A Rule for the canonical truths and a logical definition for necessity

Now, we have the half of what we need in order to formulate necessity according to the Leibnizian thesis. The other half we need is a logical representation of the canonical truths. It is quite easy. In our dialogical semantics the complex formula can be challenged according to the particle rules, and an atomic formula is not challengeable but the proponent can not assert them unless the opponent has asserted them before. Now the canonical truths according to their nature are agreed upon and thus unchallengeable. Let us introduce a new constant to represent the set of canonical truths: "+". Such a constant should be treated as an atomic formula in the course of a dialogue since no attack is allowed against it. However it has the special feature that the proponent can assert it every where in the dialogue, if according to other rules he needs such an assertion, without it necessarily having been asserted by the opponent. Therefore, the semantic rule concerning + is this:

> *Both the proponent and the opponent are permitted to assert +. It is treated like an elementary assertion namely no attack against it is possible. However, the proponent may assert it even without it being asserted before by the opponent.*

Now, as it was our intention, we define necessity as follows:

$$\Box p \equiv (+ \prec p) \tag{4.13}$$

There is another constant which may seem similar to our new constant +, i.e. ⊤. However in regard to their meanings they are

4.3. NECESSITY AND STRICT IMPLICATION 315

different, though the logical rule corresponding to them are the same and they are logically equivalent. ⊤ may read "there is a truth", while + is the set of all canonical truths. It is easy to see that we have both + → ⊤ and ⊤ → +, so as I said they are logically equivalent. However in regard to the dialogical framework their meanings are different. As far as we perform a *formal* dialogue they may be used technically interchangeably, but in a *material* dialogue which is to examine a logical consequence of some given premises in a specific context, + must be declared as the basic truths of the field, beside other truths that are given as premises but not as necessary. Imagine a thought experiment about a physical problem; we set an assumptive situation then depart to examine a thesis. Here, beside the possible facts assumed, we should observe some principles as unchallengeable. So according to our rule these latter judgments can be recalled by the proponent everywhere in the dialogue, while other truths can not be recalled under an attack toward a strict implication.

The fact that in a particular dialogue, or a particular argumentation, in this or that field, the participants only declare some fundamental truth of that field, namely some part of + whose content concern the fundamental facts of that field, should not be understood as if we have different kinds of + each for every particular field. Rather, + is universal and it can be used every where, though if it would be required to be explicit about it before entering a dialogue, it is suitable to just mention those parts which would be related to the thesis under discussion.

In relation to the mentioned point, I shall emphasize that the definition of necessity suggested here has nothing to do with the idea of *relative necessity* which is known in the relevant literature. Although our formula of definition may seem to be similar to the definition of relative necessity introduced by Smiley (Smiley, 1963), we should notice that both in regard to the motivation behind them and to their proper meanings, they are essentially different. First of all, Smiley just take as primitive a notion of absolute necessity and then define relative necessity on the basis of it. I do not want here mention the possible philosophical objections against such an idea nor the formal problems of the kinds that are shown by Humberstone (Humberstone, 1981). However, it would be worth mentioning that Humberstone's own alternative is based on the possible world semantics, then again the philosophical conception he develops is not that much different from the customary one.

Another thing which perhaps I should stress concerns the notion of absolute necessity. For those familiar with the concept of relative necessity, absolute necessity is taken to be equal with the logical necessity. This is not the case for the notion of necessity as understood in this project. To explain, recall the difference between a physical *truth* and a logical *truth*. The fact that we have a logic in which we can also deal with a physical truths does not convert them to logical truths. For a truth to be logical means that it is obtainable only according to the logical rules. The case is the same for necessity. We introduce a logic to deal with necessity, but this does not mean that the necessity at work is logical necessity. If we have $\vdash p$, it can be shown that we would have $\Box p$ and this is a logical necessity. But it is not generally true that $\Box p$ would imply that $\vdash p$. In other words, not every necessity is logical, though it is still a necessity, namely not a relative but absolute necessity.

Now let us go back to the discussion on the consequences of the introduced account of necessity.

It is easy to show that necessity entails truth, namely the formula **T** holds:

	O		P	
			$(+ \,\text{-}3\, p) \to p$	(0)
(1@0)	$+ \,\text{-}3\, p$		$+$	(2@1)
(3ⓡ2)	p		p	(4ⓡ1)

4.3. NECESSITY AND STRICT IMPLICATION

In the above dialogue, **P** is able to assert + in the move 2 in order to counterattack **O**'s move 1; then she has no choice but to assert p and so the proponent use it to respond the first attack. There is no other move possible and **P** wins.

As it is already known strict implication should be equal to necessity of a material implication. That is:

$$(p \strictif q) \equiv (+ \strictif (p \to q)) \tag{4.14}$$

Or:

$$(p \strictif q) \equiv \Box(p \to q) \tag{4.15}$$

The formula 4.15 is normally considered as the definition of the strict implication. But in the present approach, the strict implication has priority over necessity. And necessity is defined on the basis of the strict implication as stated in the formula (6). Therefore the formula 4.15, or its equivalent the formula 4.14, is not a definition and needs a proof if it is true. Indeed it is true and the dialogues to demonstrate the validity of both cases of the biconditional are brought in tables 6 and 7.

It can be shown that the modal system so defined is close to **S4**, since it validates the principles of **K** and **S4**. The validity of these formulas are demonstrated in tables 8 and 9.

$$\mathbf{K} : \Box(p \to q) \to (\Box p \to \Box q)$$

$$\mathbf{S4} : \Box p \to \Box\Box p$$

In order to show that the presented semantics is equal to **S4** one should prove the completeness metatheorem, but my aim here was not to provide a semantics for an existing system, rather it is to introduce a new approach capable to be employed in order to explain the logical nature of necessity. Moreover, I just focused on necessity; and although possibility may be defined in an easy way as ¬□¬, it is not so favorable according to the attitude that has motivated the current investigation. I think that possibility has its

Table 6:

O		P	
		$(p \mathrel{\text{⊰}} q) \rightarrow (\vdash \mathrel{\text{⊰}} (p \rightarrow q))$	(0)
(1@0)	$p \mathrel{\text{⊰}} q$	$\vdash \mathrel{\text{⊰}} (p \rightarrow q)$	(2⊤1)
(3@2)	$\vdash$	$p \rightarrow q$	(4⊤3)
(5@4)	p	p	(6@1)
(7⊤6)	q	q	(8⊤5)

Table 7:

O		P	
		$(\vdash \mathrel{\text{⊰}} (p \rightarrow q)) \rightarrow (p \mathrel{\text{⊰}} q)$	(0)
(1@0)	$\vdash \mathrel{\text{⊰}} (p \rightarrow q)$	$p \mathrel{\text{⊰}} q$	(2⊤1)
(3@2)	p	$\vdash$	(4@1)
(5⊤4)	$p \rightarrow q$	p	(6@5)
(7⊤6)	q	q	(8⊤3)

4.3. NECESSITY AND STRICT IMPLICATION

Table 8:

	O		P	
			$(+\beth(p \to q)) \to ((+\beth p) \to (+\beth q))$	(0)
(1@0)	$+\beth(p \to q)$		$(+\beth p) \to (+\beth q)$	(2r1)
(3@2)	$+\beth p$		$+\beth q$	(4r2)
(5@4)	$+$		$+$	(6@1)
(7r6)	$p \to q$		p	(8@3)
(9r8)	p			(10@7)
(11r10)	q		q	(12@5)

Table 9:

	O		P	
			$(+\beth p) \to (+\beth(+\beth p))$	(0)
(1@0)	$+\beth p$		$+\beth(+\beth p)$	(2r1)
(3@2)	$+$		$+\beth p$	(4r3)
(5@4)	$+$		$+$	(6@1)
(7r6)	p		p	(8r5)

own independent meaning and at least there are some significant notions recalled by the term "possibility" which are not able to be indirectly explained in term of necessity.[40]

However, if we just define possibility as $\neg\Box\neg$, what would be given is the classical version of **S4**, not **S5**. In the standard classical version of **S4** we have the theorem $\Diamond\Diamond p \to \Diamond p$, beside the theorem $p \to \Diamond p$. In the following I bring dialogues to demonstrate the validity of the latter in our semantic framework, once one choses to define $\Diamond$ as $\neg\Box\neg$. The validity of the formula $\Diamond\Diamond p \to \Diamond p$ is shown in table 10.

	O		**P**	
			$p \to \neg(+\ \text{-3}\ \neg p)$	(0)
(1@0)	p		$\neg(+\ \text{-3}\ \neg p)$	(2ⓡ1)
(3@2)	$+\ \text{-3}\ \neg p$		$+$	(4@3)
(5ⓡ4)	$\neg p$		p	(6@5)
	———			

The dialogue in table 11 is brought to show why the present dialogue semantics does not validate the **5** axiom: $\Diamond p \to \Box\Diamond p$, which, according to the above definition, would be equal to $(\neg(+\ \text{-3}\ \neg p)) \to (+\ \text{-3}\ (\neg(+\ \text{-3}\ \neg p)))$.

There **P** has no further move because he cannot attack the move 1 (he obviously cannot attack the move 7 since it requires to assert an elementary formula which has not been asserted by **O**) since he is under the attack toward a strict implication posed in the move 3 then he would be able to refer to the move 1 only if the main

---

[40]For instance, under the term "possible" we may intend the notion "conceivable" or "intelligible"; also we may use it as for the notion of "imaginable". Yet imaginability and conceivability can be considered as having their own logical charactristics alongside with that of possibility; and in neither case there is no such a straightforward definition on the basis of necessity. A fruitful discussion about the conceptual difference between possibility, imagination and conception can be found in (Beziau, 2016).

4.3. NECESSITY AND STRICT IMPLICATION

Table 10:

	O	P	
		$\neg(+\ \Im\ \neg(\neg(+)\neg\ \Im\ \neg(+\ \Im\ \neg p))) \to \neg(+\ \Im\ \neg p)$	(0)
(1@0)	$\neg(+)\neg\ \Im\ \neg(+\ \Im\ \neg p))$	$\neg(+)\neg\ \Im\ \neg p$	(2r1)
(3@2)	$+\ \Im\ \neg p$	$+\ \Im\ \neg(\neg(+\ \Im\ \neg p))$	(4@1)
(5@4)	$+$	$\neg(\neg(+\ \Im\ \neg p))$	(6r5)
(7@6)	$\neg(+)\neg\ \Im\ \neg p$	$+\ \Im\ \neg p$	(8@7)
(9@8)	$+$	$\neg p$	(10r9)
(11@10)	p	$+$	(12r3)
(13r12)	$\neg p$	p	(14@13)

Table 11:

	O	P	
		$\neg(\neg(+\ \Im\ \neg p)) \to (+\ \Im\ \neg(\neg(+\ \Im\ \neg p)))$	(0)
(1@0)	$\neg(+\ \Im\ \neg p)$	$+\ \Im\ \neg(\neg(+\ \Im\ \neg p))$	(2r1)
(3@2)	$+$	$\neg(+)\neg\ \Im\ \neg p$	(4r3)
(5@4)	$+\ \Im\ \neg p$	$+$	(6@5)
(7r6)	$\neg p$		

connective of that move would be strict implication but now it is not.

The meaning of necessity explained in the discussed dialogical way would be same for both classical and intuitioinistic logic; and it is, as always, the difference between a structural rule of them which causes the differences. Therefore the problem for the interpretation of intuitionistic modal logic which arises in the possible world semantics has no place here. By choosing (SR-5i) instead of (SR-5c) we may obtain intuitionistic version of, say, **S4**. One can use the rules for quantifier and develop a first-order modal logic—but I drop that job here and postpone it to further works, since the main objective of the present discussion is to sketch out the main idea of an alternative way of explanation of the logical behavior of necessity.

A suggestion for the possibility

However, if we want to define possibility on the basis of necessity, which may correspond to the third conception listed in page 299, I would prefer to defines $\Diamond$ as $\sim \Box \neg$ rather than $\neg \Box \neg$. It means that to say that A is possible, in this sense, is that there is no evidence for A being necessarily absurd. According to the customary definition to say that A is possible is equal to say that it is absurd that A's strong negation be necessary. Such a definition seems stronger than what we may intend by considering something as possible. Indeed, in the course of reasoning when we consider something as possible, in the mentioned conception, we just say that there is no evidence for it's being necessarily wrong, in the sense that it is permissible, according to our current state of knowledge, to *pose* it during the argumentation.

If we define $\Diamond$ as $\sim \Box \neg$, then, while working in the intuitionistic framework we will have $\Diamond p \vee \Box \neg p$ and $\neg \Diamond p \to \Box \neg p$ which are good features. We will still have the essential feature $p \to \Diamond p$, which can be shown easily. However we will not have $\Diamond \Diamond p \to \Diamond p$.

4.3.4 Analysis: modality and dialogue

To summarize, in order to formulate necessity we appeal to the notions of fundamental truths and that of strict implication. Fundamental truths and those are strictly implied by them are necessary.

A necessary truth is not contingent and in this sense it is independent of the perceptual experience. However, we are not to say that a necessary truth is always a formal or logical truth. The fundamental truths may not be, and mostly they are not, analytical—in the Kantian sense. For example that 2 is greater than 1 is a fundamental truth, thus necessary, in the field of mathematics. However, it is synthetic in the sense that it not result of a mere logical proof. In any case it should be *a priori*, namely it is not the subject of a perceptual examination. Nevertheless, this does not mean that they are independent of the experience in general, for they may be discovered or obtained by intuition which is by no means immediate.[41] Indeed most of *a priori* truths are gradually discovered as a science progresses. In that sense, as I have stated before, the content of + can be enriched in the course of time.

So we should be careful not to equate necessity with logical validity or analyticity. Of course every logical truth is necessary. Formally speaking the genuine logical axioms should be included in + and since from $\psi \vdash \phi$ we have $\psi \prec \phi$, every logical theorem is strictly implied by the axioms and thus is necessary.[42] But the realm of necessity is not restricted to the logical validity.

---

[41]Notice that Leibniz's use of the word analytic is different from that of Kant and that in our time. Leibniz uses this word in order to indicate that a truth is not result of a factual coincidence but it is intrinsic in its subject. In this sense also Kant would admit that, for example, 5+7=12 is analytic. But Kant uses the term analytic when we just work with formal definitions and formal relations. So he considers 5+7=12 as synthetic for we deal with the contents, although he admits that the truth of the mentioned claim rests only on the contents of its subjects. We follow Kantian terminology and say, in a complete agreement with Leibnizian attitude, that there are synthetic necessities.

[42]Provided that the axioms are taken to be necessary, or there is no axiom.

However, what is important is the role of necessity in the dialogue. We may understand this modality as a way we deal with a truth so modified in the course of dialogue. The explanation goes as follows:

Imagine that we are to argue about a thesis. There are a set of premises which are agreed upon by the participants. However, we have two kinds of premises, a set of fundamental principles governing the field about which we are arguing, and a set of assumptions which are relevant to our current discussion. Now assume during the dialogue someone claims that A is necessary, and then the other asserts "O.k. show it!". Here, the defender may no longer use the assertions made before this claim nor the assumptions, however he/she can use the fundamental principles since they are supposed to constitute all knowledge belonging to the field. Then necessity is conceived in relation to argumentation, necessary truths are those that can be recalled under any assumption, and can be proved under any assumption.[43] While contingent truths can be suspended under some assumptions, and they should be so, if we want to examine the essential properties of the objective domains.

The rationale behind the two rules will be clear by considering a purely knowledge oriented dialogue. While the implication is supposed to be treated in the way that if someone asserts an implication he/she is obliged to assert the consequence if the other party gives the antecedent, now in the strict implication one may use only what are asserted after the assertion of the antecedent, namely any assumption which is asserted before the challenge should be suspended. However if the challenger has asserted some claims as unsuspendable, namely as necessary, the defender is allowed to recall them if needed. Accordingly if one claims that a judgment is necessary, one should be able to show it only using the fundamental truths and the claims that are accepted as necessary by the challenger. Following such an idea and formulating the rules

---

[43]One may say they are unsuspendable and indeed serve as the ground in an objectively oriented argumentation.

to represent them we have obtain a semantic framework to deal with a logic containing necessity without committing to a specific metaphysical viewpoint or committing to reducing the notion of necessity and loosing its possible primitive meaning.[44]

The dialogical logic provides us with a tool to understand the meaning of necessity and also to formulate it within a logical system. The PWS is not the indispensable way of interpretation of necessity, or other modalities; and the metaphysical problems arisen within it are not the genuine problems pertaining to the modalities themselves.

Here I just discussed the issue of necessity for the propositional logic. Nevertheless, it is plausible to follow the presented line of thoughts to formulate "possibility" (if there is an independent notion of it), first-order modal logic and the relation of identity. The key point is that instead of simulating a metaphysical model we begin with the role of the notions in question in the dialogue. Of course this is not to say that the PWS is flawed, rather that we should not overestimate its philosophical power, however technically fruitful it is. When the objective is the philosophical explanations, dialogical method is more appropriate to represent the essence of the logical issues and it is of less metaphysical presuppositions.

4.4 Conclusion

As it is discussed in the previous chapters, picture theory of meaning, and also truth-conditional account of meaning, are inappropriate according to the phenomenological investigation. Accordingly, the transcendental phenomenology refutes any kind of extensional approach, including model-theoretical interpretation, to constitute the meaning of logical entities. Therefore, truth-functional definitions,

---

[44]I am not saying that logic should be metaphysically neutral, rather I think that metaphysics should be studied in its own right and correlatively this would have some consequences for logic. It is by no means acceptable that a technical tool for a logic forces a specific metaphysics.

though fruitful in algebra and informatics, are not capable to explain the proper meaning of the logical constants. Use theory and inferentialism have their own shortcomings which are discussed in the previous investigations. In the present chapter I tried to elaborate in a more technical manner the phenomenological meaning explanation of some logical connectives by referring to the intentionality constituting them. I tried to show that dialogical framework is capable to both contribute to this meaning explanation and also to formalize the characteristics of the logical constants in a way phenomenologically examinable, and if suitable, admissible. Of course in the present project I deal with the questions concerning the meaning, so I dealt with the particle rules and I leave the discussions about the structural rules—which belong, not to the meaning explanation but, to the truth level. So, I did not go thorough in realizing the phenomenological project of revising logic and developing a unified pure logic, though I tried to show that it is in principle feasible.

As it is the case for every meaning, the meanings of the logical constants should be explained first of all in terms of intentionality. However, such an intentionality in the case of logical issues can be properly explored within the dialogical framework. In the present chapter I tried to carry out this meaning explanation for negations, a kind of implication and necessity. With further investigation we may progress in explaining the meanings of other logical constants. Provided with the method of dialogical logic, the phenomenological theory of meaning, is able also to clearly explain, and make precise as well, the meaning of logical connectives.

Bibliography

K. Apel. *Towards a Transformation of Philosophy*, trl. G. Adey and D. Fisby. Marquette University Press, Milwaukee, 1998.

R. E. Auxier and L. E. Hahn. *The Philosophy of Michael Dummett*. Library of Living Philosophers, Vol. 31. Open Court, 2007.

S. Bachelard. *A Study of Husserl's Formal and Transcendental Logic*, trl. L. E. Embree. Northwestern University Press, Evanston, 1968.

G.P. Baker and P.M.S. Hacker. *Scepticism, Rules and Language*. Blackwell, Oxford, 1984.

G.P. Baker and P.M.S. Hacker. *Wittgenstein: Rules, Grammar and Necessity, Vol. 2 of an Analytical Commentary on the "Philosophical Investigations"*. Blackwell, Oxford, 1985.

C. Beyer. Edmund Husserl. In E. Zalta, editor, *The Stanford Encyclopedia of Philosophy*. Summer 2015 edition, 2015.

J.-Y. Beziau. Possibility, imagination and conception. *Principios*, 23(40):59–95, 2016.

M. Brainard. *Belief and its Neutralization: Husserl's System of Phenomenology in Ideas I*. State University of New York Press, Albany NY, 2002.

L. E. J. Brouwer. Essentieel negatieve eigenschappen [essentially negative properties]. *Indagationes Mathematicae*, 10:322–323, 1948. English translation in (Brouwer, 1975, pp. 478–9).

L. E. J. Brouwer. The effect of intuitionism on classical algebra of logic. *Proceedings of the Royal Irish Academy*, 57:113–116, 1955. Reprinted in (Brouwer, 1975, pp. 551–4).

L. E. J. Brouwer. *Collected Works, I.* ed. A. Heyting. North-Holland, Amsterdam, 1975.

L. E. J. Brouwer. Life, art, and mysticism, trl. W. van Stigt. *Notre Dame Journal of Formal Logic*, 37/3:389–429, 1996.

M. Budd. *Wittgenstein's Philosophy of Psychology.* Routledge, London, 1989.

K. Bühler. *Theory of Language: The representational function of language.* John Benjamins Publishing Company, Amsterdam / Philadelphia, 2011.

S. Candlish and G. Wrisley. Private language. In E. Zalta, editor, *The Stanford Encyclopedia of Philosophy.* 2012.

H. Capell and E. Niño. Belief and knowledge in Gustave Le Bon and José Ortega y Gasset. *Revista de historia de la psicología*, 19: 517–522, 1998.

N. Clerbout. Finiteness of plays and the dialogical problem of decidability. *IfCoLog Journal of Logics and their Applications*, 1:1:115–130, 2014.

J. Derrida. *Speech and Phenomena,* trl. D. B. Allison. Northwestern University Press, 1973.

J. Derrida. *Edmund Husserl's Origin of Geometry: An Introduction,* trl. J. P. Leavey. University of Nebraska Press, 1989.

J. Dodd. The depth of signs. *Graduate Faculty Philosophy Journal*, 33(1):3–26, 2012.

K. Došen. *Logical Constants: An Essay in Proof Theory*, D. Phil thesis, University of Oxford. 1980.

M. Dummett. *The Logical Basis of Metaphysics*. London, Duckworth, 1991.

M. Dummett. *Origins of Analytical Philosophy*. Harvard University Press, Cambridge,Massachusetts, 1994.

M. Dummett. *Elements of Intuitionism*. Oxford Logic Guides 39. Oxford University Press, USA, 2nd edition, 2000.

M. Farber. *The Foundation of Phenomenology*. Aldine Transaction, London, 2006.

W. Felscher. Dialogues as a foundation for intuitionistic logic. *Handbook of Philosophical Logic,* ed. D. Gabbay and F. Guenthner, III:341–72, 1986.

G. Frege and E. Husserl. *Correspondance Frege-Husserl.* 1987.

P. Geach. *Mental Acts: their Content and their Objects.* Routledge and K. Paul, London, 1971.

H. Glock. *A Wittgenstein Dictionary.* Blackwell, Oxford, 1996.

K. Gödel. Zum intuitionistischen Aussagenkalkül. *Anzeiger der Akademie der Wissenschaften in Wien,* 69:65–66, 1932. Also, with English translation, in (Gödel, 1986, pp. 222–225).

K. Gödel. *Collected Works. I: Publications 1929–1936.* Oxford University Press, Oxford, 1986.

J. Habermas. An alternative way out of the philosophy of the subject: Communicative versus subject-centered reason. In *From Modernism to Postmodernism: An Anthology,* L. Cahoone (ed.), pages 589–616. Blackwell, Oxford, 1996.

P. Hacker. *Appearance and Reality.* Oxford: Blackwell, 1987.

M. Hartimo. Husserl and the algebra of logic: Husserl's 1896 lectures. *Axiomathes*, 22(1):121–133, 2012.

A. Hermann. *To Think Like God: Pythagoras and Parmenides. The Origins of Philosophy.* Parmenides Publishing, Las Vegas, 2004.

A. Heyting. Sur la logique intuitionniste. *Académie Royale de Belgique, Bulletin de la Classe des Sciences*, 16:957–963, 1930.

A. Heyting. Griss and his Negationless Intuitionistic Mathematics. *Synthese*, 9(1):91 – 96, 1955.

J. Hintikka. *On Wittgenstein.* Thomson Learning, Wadsworth, 2000.

B. Hopkins. *The Origin of the Logic of Symbolic Mathematics. Edmund Husserl and Jacob Klein.* Bloomington: Indiana University Press, 2011.

W. Hopp. Husserl, Dummett, and the linguistic turn. *Grazer Philosophische Studien*, 78(1):17–40, 2009.

L. Humberstone. Relative necessity revisited. *Reports on Mathematical Logic*, 13:33–42, 1981.

E. Husserl. *Logische Untersuchungen, Zwiter Band, I. Teil.* Max Niemeyer, Tübingen, 1913.

E. Husserl. *Erfahrung und Urteil,* ed. L. Landgrebe. Academie Verlagsbuchhandlung, Prag, 1939.

E. Husserl. *Ideen zu einer reinen Phänomenologie und phänomenologischen Philosophie: Zweites Buch: Phänomenologische Untersuchungen zur Konstitution.* Husserliana: Edmund Husserl — Gesammelte Werke, Band 4. Martin Nijhoff, The Hague, Netherlands, 1952.

E. Husserl. *Die Krisis der europäischen Wissenschaften und die transzendentale Phänomenologie,* trl. F. Kersten, volume VI of *Husserliana.* Martin Nijhoff, The Hague, Netherlands, 1954.

E. Husserl. *Erste Philosophie (1923/4). Zweiter Teil: Theorie der phänomenologischen Reduktion, Husserlina IIX* Edited by Rudolf Boehm. Martinus Nijhoff, The Hague, Netherlands, 1959.

E. Husserl. *Cartesian Meditations,* trl. Dorian Cairns. Springer Science+Business Media, Dordrecht, 1960.

E. Husserl. *Logische Untersuchungen, Zwiter Band, II. Teil.* Max Niemeyer, Tübingen, 1968.

E. Husserl. *Formal and Transcendental Logic,* trl. D. Cairns. Martin Nijhoff, The Hague, Netherlands, 1969.

E. Husserl. *Ideas: General Introduction to Pure Phenomenology,* trl. W. R. Gibson. George Allen & Unwin Ltd, London, 1969.

E. Husserl. *Philosophie der Arithmetik,* Husserliana, Band XII. Martinus Nijhoff, The Hague, Netherlands, 1970.

E. Husserl. *Experience and Judgment,* trl. J. Churchill, and K. Ameriks. Routledge, London, 1973.

E. Husserl. *Formale und Transzendentale Logik,* ed. P. Janssen, volume XVII of *Husserliana.* Martin Nijhoff, The Hague, Netherlands, 1974.

E. Husserl. *Ideen zu einer reinen Phänomenologie und phänomenologischen Philosophie. Erstes Buch. 1. Halbband: Text der 1.-3. Auflage.*, volume III/1 of *Husserliana.* Martin Nijhoff, The Hague, Netherlands, 1976.

E. Husserl. *Einleitung in die Logik und Erkenntnistheorie Vorlesungen 1906/07* ed. U. Melle, volume XXIV of *Husserliana.* Springer Netherlands, 1985.

E. Husserl. *Vorlesungen über Bedeutungslehre Sommersemester 1908*, volume XXVI of *Husserliana*. Springer Netherlands, 1986.

E. Husserl. *Ideas II*. 1988. translated by R. Rojcewicz and A. Schuwer.

E. Husserl. Phänomenologie und Anthroplogie. *Aufsätze und Vorträge. 1922-1937, Husserliana XXVII*, Edited by T. Nenon and H.R. Sepp, pages 164–81, 1988.

E. Husserl. *Formale und Transzendentale Logik*. Felix Meiner Verlag, Hamburg, 1992.

E. Husserl. A review of volume i of ernst schröder's vorlesungen über die algebra der logik. In D. Willard, editor, *Early Writings in the Philosophy of Logic and Mathematics*, volume 5 of *Edmund Husserl Collected Works*, pages 52–91. Springer Science+Business Media, Dordrecht, 1994a.

E. Husserl. The deductive calculus and the logic of contents. In D. Willard, editor, *Early Writings in the Philosophy of Logic and Mathematics*, volume 5 of *Edmund Husserl Collected Works*, pages 92–120. Springer Science+Business Media, Dordrecht, 1994b.

E. Husserl. *Analyses Concerning Passive and Active Syntheses*, trl. A. J. Steinbock. Martin Nijhoff, The Hague, Netherlands, 2001.

E. Husserl. *Logik: Vorlesung 1896*, ed. Elisabeth Schuhmann. Husserliana: Edmund Husserl — Materialien, Band 1. Springer Netherlands, 1 edition, 2001a.

E. Husserl. *Logical Investigations (2 vols.)* trl. J. Findlay, ed. D. Moran. Routledge, London, 2001b.

E. Husserl. *Logische Untersuchungen. Ergänzungsband. Erster Teil* ed. U. Melle, volume XX/1 of *Husserliana*. The Hague, Netherlands, 2002.

E. Husserl. *Philosophy of Arithmetic,* trl. D. Willard. Kluwer Academic Publishers, Dordrecht, 2003.

E. Husserl. *Logische Untersuchungen. Ergänzungsband. Zweiter Teil* ed. U. Melle, volume XX/2 of *Husserliana.* The Hague, Netherlands, 2005.

E. Husserl. *Introduction to Logic and Theory of Knowledge: Lectures 1906/07* trl. C. O. Hill. Edmund Husserl — Collected Works, Vol. 13. Springer, Dordrecht, 2008.

E. Husserl. *Grenzprobleme der Phänomenologie,* ed. R. Sowa and T. Vongehr, volume 42 of *Husserliana.* Springer Netherlands, 2014.

P. Johnston. *Wittgenstein: Rethinking the Inner.* Routledge, London, 1993.

K. Kaneiwa. On the semantics of classical first-order logic with constructive double negation. In *IICAI,* pages 1225–1242, 2005a.

K. Kaneiwa. Negations in description logic-contraries, contradictories, and subcontraries. *Contributions to ICCS,* pages 66–79, 2005b.

A. Kenny. *Wittgenstein.* Blackwell, Oxford, 2006.

G. Kreisel. Informal rigour and completeness proofs. In I. Lakatos, editor, *Problems in the Philosophy of Mathematics,* page 138–186. North-Holland, Amsterdam, 1967.

S. Kripke. *Wittgenstein on Rules and Private Language.* Blackwell, Oxford, 1982.

V. A. Ladov. The discussion on the private language argument: Linguist versus philosopher. *Studia Humana,* 1/1:10–23, 2012.

S. Law. Five private language arguments. *International Journal of Philosophical Studies,* 12/2:152–176, 2004.

G. Leibniz. *Philosophical Papers and Letters*, Ed. L. E. Loemker. Kluwer Academic Publishers, Dordrecht, 1989.

E. Levinas. *Totality and Infinity: An Essay on Exteriority*. Martinus Nijhoff Philosophy Texts. Springer, 4th edition, 1979.

D. Lewis. *Counterfactuals*. Blackwell Publishers, Oxford, 1973.

D. Lohmar. Beiträge zu einer phänomenologischen Theorie des negativen Urteils. *Husserl Studies*, 8(3):173–204, 1992.

D. Lohmar. Elements of a phenomenological justification of logical principles, including an appendix with mathematical doubts concerning some proofs of Cantor on the transfiniteness of the set of real numbers. *Philosophia Mathematica*, 10(2):227–250, 2002a.

D. Lohmar. Husserl's concept of categorial intuition. In *One Hundred Years of Phenomenology: Husserl's Logical Investigations Revisited*, pages 125–145. Springer Netherlands, Dordrecht, 2002b. Edited by D. Zahavi and F. Stjernfelt.

D. Lohmar. The transition of the principle of excluded middle from a principle of logic to an axiom. *New Yearbook for Phenomenology and Phenomenological Philosophy*, 4:53–68, 2004.

D. Lohmar. Categorial intuition. In *A Companion to Phenomenology and Existentialism*, pages 115–126. Blackwell, Oxford, 2006. Edited by H. Dreyfus, M. Wrathal.

S. Luft. *Phänomenologie der Phänomenologie: Systematik und Methodologie der Phänomenologie in der Auseinandersetzung zwischen Husserl und Fink*. Phaenomenologica 166. Springer Netherlands, 2002.

N. Malcolm. Wittgenstein on language and rules. *Philosophy*, 64: 5–28, 1998.

M. Marion. Game semantics and the manifestation thesis. In *The Realism-Antirealism Debate in the Age of Alternative Logics,* eds. S. Rahman, G. Primiero and M. Marion, Logic, Epistemology, and the Unity of Science, pages 141–168. 2012.

M. McGinn. *Wittgenstein and the Philosophical Investigations.* Routledge, London, 1997.

R. McKirahan. *Philosophy Before Socrates.* Hackett Publishing Company, Indianapolis, 2011.

U. Meixner. *Defending Husserl: A Plea in the Case of Wittgenstein and Company Versus Phenomenology.* De Gruyter, Berlin, 2014.

U. Melle. Signitive und signifikative Intentionen. *Husserl Studies,* 15(3):167–181, 1998.

J. Mensch. Derrida–Husserl: Toward a phenomenology of language. *New Yearbook for Phenomenology and Phenomenological Philosophy,* pages 1–66, 2001.

G. Mints. Notes on constructive negation. *Synthese,* 148(3):701 – 717, 2006.

J. N. Mohanty. *Edmond Husserl's theory of meaning.* Martin Nijhoff, The Hague, Netherlands, 1976.

B. Myers-Schulz and E. Schwitzgebel. Knowing that p without believing that p. *Noûs,* 47(2):371–384, 2013.

J. Palmer. Parmenides. In E. Zalta, editor, *The Stanford Encyclopedia of Philosophy.* 2012.

D. Pears. *The False Prison, Vol2.* Blackwell, Oxford, 1996.

C. S. Peirce. The fixation of belief. *Popular Science Monthly,* 12: 1–15, 1877.

J. Piaget. *La Representation du Monde chez L'enfant*. Presses universitaires de France, Paris, 1947.

A. Pietarinen. Two papers on existential graphs by Charles Peirce. *Synthese*, 192(4):881–922, 2015.

K. Puhl and S. Rinofner-Kreidl. Is every mentalism a kind of psychologism? Michael Dummett's critique of Edmund Husserl and Gareth Evans. *Grazer Philosophische Studien*, 55:213–237, 1998.

S. Rahman and L. Keiff. On how to be a dialogician. In Daniel Vanderveken, editor, *Logic, Thought and Action*, volume 2 of *Logic, Epistemology, and the Unity of Science*, chapter 17, pages 359–408. Springer Netherlands, 2005.

S. Rahman and H. Rückert. Dialogische modallogik (für T, B, S4, und S5). *Logique et Analyse*, 167(168):243–282, 1999.

H. Rückert. Why dialogical logic? In H. Wansing, editor, *Essays on Non-Classical Logic*, pages 165–185. River Edge: World Scientific, 2001.

S. Read. Harmony and modality. In C. Dégremont, L. Kieff, and H. Rückert, editors, *Dialogues, Logics and Other Strange Things: Essays in Honour of Shahid Rahman*, pages 285–303. College Publications, London, 2008.

S. Read. Necessary truth and proof. *Kriterion: Revista de Filosofia*, 51(121), 2010.

M. Rebuschi. Implicit versus explicit knowledge in dialogical logic. In O. Majer, A-V. Pietarinen, and T. Tulenheimo, editors, *Games: Unifying Logic, Language, and Philosophy*, pages 229–246. Springer, Dordrecht, 2009.

H. Remplein. *Die seelische Entwicklung des Menschen im Kindes und Jugendalter*. Reinhardt Verlag, Munich, 1966.

B. Russell. *The Principles of Mathematics*. Allen and Unwin, London, 1903.

T. Smiley. Relative necessity. *Journal of Symbolic Logic*, 28(2): 113–134, 1963.

R. Sokolowski. *Husserlian Meditations*. Northwestern University Press, Evanston, 1974.

R. Sokolowski. Husserl's concept of categorial intuition. *Philosophical Topics*, 12(Supplement):127–141, 1981.

R. Sokolowski. *Introduction to Phenomenology*. Cambridge University Press, Cambridge, 2000.

R. Spitz. *Nein und Ja, Die Ursprünge der menschlichen Kommunikation*. Klett-Cotta Verlag, Stuttgart, 1992.

L. Stone and J. Church. *Childhood and Adolescence*. Random House, New York, 1957.

S. Strasser. *The Idea of Dialogal Phenomenology*, trl. H. Koren. Duqusne University Press, Pittsburgh, 1969.

G. Sundholm. What is an expression? In *Logica Yearbook 2001*, pages 181–194. 2002.

G. Sundholm and M. van Atten. The proper explanation of intuitionistic logic: On Brouwer's demonstration of the Bar Theorem. In *One Hundred Years of Intuitionism (1907-2007)*, eds. M. van Atten, P. Boldini, M. Bourdeau and G. Heinzmann , pages 60–77. Birkhäuser, Bâle, Boston, Berlin, 2008.

R. Swinburne. *The Existence of God*. Oxford University Press, 2nd edition, 2004.

L. Taran. *Parmenides: A Text with Translation, Commentary, and Critical Essays*. Princton University Press, Princeton, New Jersey, 1965.

N. Tennant. Negation, absurdity and contrariety. In *What is Negation?* eds. Gabbay, Dov M. and Wansing, Heinrich, pages 199–222. Springer Netherlands, Dordrecht, 1999.

R. Tieszen. *Phenomenology, Logic, and the Philosophy of Mathematics*. Cambridge University Press, Cambridge, 2005.

A. Troelstra and D. van Dalen. *Constructivism in Mathematics*. North-Holland, Amsterdam, 1988.

E. Tugendhat. *Der Wahrheitsbegriff Bei Husserl Und Heidegger*. De Gruyter, Berlin, 1967.

M. van Atten. Gödel, mathematics, and possible worlds. *Axiomathes*, pages 355–363, 2001.

M. van Atten. Why Husserl should have been a strong revisionist in mathematics. *Husserl Studies*, 18(1):1–18, 2002.

M. van Atten. *On Brouwer*. Wadsworth, Belmont, 2004.

M. van Atten. The hypothetical judgement in the history of intuitionistic logic. In C. Glymour, W. Wang, and D. Westerstahl, editors, *Logic, Methodology, and Philosophy of Science XIII: Proceedings of the 2007 international congress in Beijing*, page 122–136, London, 2009. College Publications.

M. van Atten. Construction and constitution in mathematics. *The New Yearbook for Phenomenology and Phenomenological Philosophy*, X:43–90, 2010.

M. van Atten. The development of intuitionistic logic. In E. Zalta, editor, *The Stanford encyclopedia of philosophy*. 2014.

M. van Atten. On the fulfillment of certain categorial intentions. *The New Yearbook for Phenomenology and Phenomenological Philosophy*, 13:173–185, 2015.

M. van Atten and G. Sundholm. L.E.J. Brouwer's 'Unreliability of the logical principles': A new translation, with an introduction. *History and Philosophy of Logic*, 38(1):24–47, 2016.

D. van Dalen. The role of language and logic in Brouwer's work. In *Logic in Action* ed. E. Orlovska, pages 3–14. Springer, Vienna, 1999.

D. van Dalen. *The Selected Correspondence of L.E.J. Brouwer*. Springer, London, 2011.

A. Varzi and M. Warglien. The geometry of negation. *Journal of Applied Non-Classical Logics*, 13(1):9–19, 2003.

H. Wang. *A Logical Journey. From Gödel to Philosophy*. MIT Press, Cambridge, 1996.

D. Whiting. Truth: The aim and norm of belief. *Teorema: International Journal of Philosophy*, 32(3):121–136, 2013.

T. Williamson. Modal logic within counterfactual logic. In B. Hale and A. Hoffmann, editors, *Modality: Metaphysics, Logic, and Epistemology*. OUP Oxford, 2010.

P. Winch. *The Idea of a Social Science and its Relation to Philosophy*. Routledge, London, 1990.

L. Wittgenstein. *On Certainty* Ed. Anscombe and von Wright. Harper Torchbooks, Oxford, 1969.

L. Wittgenstein. *Letters to Russell, Keynes and Moore,* edited by G. von Wright and B. McGuinness. Cornell University Press, Ithaca, 1974.

L. Wittgenstein. *Philosophical Investigations,* The German Text with an English Translation by G. E. M. Anscombe, P. M. S. Hacker and Joachim Schulte. Blackwell, Oxford, 2009.

L. Wittgenstein. *Tractatus Logico-Philosophicus,* The original German, alongside both the Ogden/Ramsey, and Pears/McGuinness English translations. http://people.umass.edu/klement/tlp/, 2015.

D. Zahavi. Horizontal intentionality and transcendental intersubjectivity. *Tijdschrift voor Filosofie,* 59/2:304–321, 1997.

D. Zahavi. *Husserl and Transcendental Intersubjectivity: A Response to the Linguistic-Pragmatic Critique,* trl. E. A. Behnke. Ohio University Press, Athens, 2001.

Index of citations

Apel (1998), 23, 327
Auxier and Hahn (2007), 232, 233, 327
Baker and Hacker (1984), 21, 327
Baker and Hacker (1985), 21, 327
Beyer (2015), 42, 327
Beziau (2016), 320, 327
Brainard (2002), 104, 327
Brouwer (1948), 255, 256, 327
Brouwer (1955), 259, 268, 327
Brouwer (1975), 58, 176, 256, 259, 268, 327, 328
Brouwer (1996), 58, 328
Budd (1989), 21, 328
Bühler (2011), 49, 233, 328
Bachelard (1968), 173, 327
Candlish and Wrisley (2012), 21, 25, 27, 328
Capell and Niño (1998), 155, 328
Clerbout (2014), 199, 328
Derrida (1973), 8, 28, 328
Derrida (1989), 94, 95, 328
Dodd (2012), 108, 165, 168, 328
Došen (1980), 207, 328
Dummett (1991), 215, 227, 329
Dummett (1994), 230, 329
Dummett (2000), 217–219, 221–225, 329
Farber (2006), 101, 329
Felscher (1986), 276, 329
Frege and Husserl (1987), 163, 329
Geach (1971), 23, 54, 81, 329
Glock (1996), 21, 329
Gödel (1932), 216, 329
Gödel (1986), 329
Habermas (1996), 23, 329
Hacker (1987), 20, 329
Hartimo (2012), 170, 330
Hermann (2004), 274, 330
Heyting (1930), 269, 330
Heyting (1955), 253, 255, 330
Hintikka (2000), 21, 22, 330
Hopkins (2011), 118–121, 138, 330
Hopp (2009), 235, 330
Humberstone (1981), 316, 330
Husserl (1913), 32, 39–42, 46, 47, 51–53, 60, 140, 175, 330
Husserl (1939), 45, 96, 131, 132, 211, 248, 330
Husserl (1952), 71, 129, 145, 330
Husserl (1954), 31, 94, 95, 330
Husserl (1959), 74, 191, 331
Husserl (1960), 70, 71, 75, 192,

201, 331
Husserl (1968), 98–100, 112, 122, 126, 142, 251, 331
Husserl (1969), 30, 66, 67, 92, 148, 173, 192, 331
Husserl (1970), 101, 331
Husserl (1973), 45, 128, 131, 132, 136, 211, 249, 250, 331
Husserl (1974), 123, 137, 331
Husserl (1976), 62, 67, 68, 92, 331
Husserl (1985), 228, 229, 331
Husserl (1986), 42, 50, 171, 331
Husserl (1988), 71, 128, 129, 192, 332
Husserl (1992), 30, 332
Husserl (1994a), 44, 177, 246, 332
Husserl (1994b), 177, 246, 247, 252, 332
Husserl (2001), 31, 332
Husserl (2001a), 170, 332
Husserl (2001b), 32, 38–41, 46, 47, 50, 52, 53, 55, 59, 97, 99, 100, 102, 112, 116, 118, 122, 126, 140, 142, 175, 215, 251, 252, 254, 332
Husserl (2002), 102–105, 115–117, 332
Husserl (2003), 100, 101, 164, 165, 332
Husserl (2005), 32, 49, 68, 104, 165, 230, 231, 333
Husserl (2008), 228, 229, 333

Husserl (2014), 33, 34, 333
Johnston (1993), 20, 333
Kaneiwa (2005a), 263, 333
Kaneiwa (2005b), 263, 333
Kenny (2006), 20, 21, 333
Kreisel (1967), 210, 270, 333
Kripke (1982), 21, 333
Ladov (2012), 20, 333
Law (2004), 20, 24, 333
Leibniz (1989), 300, 306, 333
Levinas (1979), 58, 334
Lewis (1973), 306, 334
Lohmar (1992), 240, 267, 334
Lohmar (2002a), 258, 259, 334
Lohmar (2002b), 119–121, 334
Lohmar (2004), 177, 240, 334
Lohmar (2006), 118, 119, 334
Luft (2002), 72, 73, 334
Malcolm (1998), 21, 334
Marion (2012), 235, 334
McGinn (1997), 21, 335
McKirahan (2011), 274, 335
Meixner (2014), 85, 335
Melle (1998), 104, 335
Mensch (2001), 28, 335
Mints (2006), 296, 335
Mohanty (1976), 33, 335
Myers-Schulz and Schwitzgebel (2013), 150, 154, 335
Palmer (2012), 274, 335
Pears (1996), 20, 335
Peirce (1877), 154, 335
Piaget (1947), 188, 335

INDEX OF CITATIONS

Pietarinen (2015), 197, 336
Puhl and Rinofner-Kreidl (1998), 235, 336
Rahman and Keiff (2005), 197, 198, 208, 283, 336
Rahman and Rückert (1999), 298, 336
Read (2008), 297, 336
Read (2010), 297, 308, 336
Rebuschi (2009), 198, 336
Remplein (1966), 188, 336
Rückert (2001), 197, 336
Russell (1903), 296, 336
Smiley (1963), 315, 337
Sokolowski (1974), 119, 337
Sokolowski (1981), 119, 121, 337
Sokolowski (2000), 119, 337
Spitz (1992), 188, 337
Stone and Church (1957), 188, 337
Strasser (1969), 180, 181, 184–189, 193–196, 337
Sundholm and van Atten (2008), 234, 337
Sundholm (2002), 42, 337
Swinburne (2004), 157, 337
Taran (1965), 274, 337
Tennant (1999), 281, 337
Tieszen (2005), 230, 338
Troelstra and van Dalen (1988), 279, 338
Tugendhat (1967), 119, 338
van Atten and Sundholm (2016), 178, 259, 338
van Atten (2001), 9, 338
van Atten (2002), 177, 338
van Atten (2004), 90, 210, 338
van Atten (2009), 239, 261, 278, 338
van Atten (2010), 137, 176, 240, 338
van Atten (2014), 268, 279, 338
van Atten (2015), 118, 121, 338
van Dalen (1999), 58, 339
van Dalen (2011), 253, 254, 339
Varzi and Warglien (2003), 264, 339
Wang (1996), 275, 339
Whiting (2013), 157, 339
Williamson (2010), 306, 339
Winch (1990), 23, 339
Wittgenstein (1969), 155, 339
Wittgenstein (1974), 85, 339
Wittgenstein (2009), 11–15, 17, 18, 80, 81, 339
Wittgenstein (2015), 43, 76, 339
Zahavi (1997), 75, 340
Zahavi (2001), 75, 340

www.ingramcontent.com/pod-product-compliance
Lightning Source LLC
Chambersburg PA
CBHW050123170426
43197CB00011B/1690